MCWP 3-11.2 w/Ch 1

Marine Rifle Squad

U.S. Marine Corps

PCN 143 000112 00

MCCDC (C 42)
27 Nov 2002

ERRATUM

to

MCWP 3-11.2

MARINE RIFLE SQUAD

1. For administrative purposes, FMFM 6-5 is reidentified as MCWP 3-11.2.

DEPARTMENT OF THE NAVY
Headquarters United States Marine Corps
Washington, D.C. 20380-0001

2 December 1991

FOREWORD

1. PURPOSE

Fleet Marine Force Manual (FMFM) 6-5, *Marine Rifle Squad*, provides basic guidance to enable the rifle platoon squad leader to fight and lead his squad in combat.

2. SCOPE

This manual describes the organization, weapons, capabilities, and limitations of the Marine rifle squad. It addresses the squad's role within the platoon and that of the fire teams within the rifle squad. Emphasis is placed on offensive and defensive tactics and techniques, as well as the different types of patrols the squad will conduct.

3. SUPERSESSION

FMFM 6-5, *Marine Rifle Squad*, dated 27 June 1988.

4. CHANGES

Recommendations for improvements to this publication are encouraged from commands as well as from individuals. Forward suggestions using the User Suggestion Form format to —

COMMANDING GENERAL
DOCTRINE DIVISION (C 42)
MARINE CORPS COMBAT DEVELOPMENT COMMAND
3300 RUSSELL ROAD
QUANTICO VA 22134-5021

5. CERTIFICATION

Reviewed and approved this date.

BY DIRECTION OF THE COMMANDANT OF THE MARINE CORPS

M. P. CAULFIELD
Major General, U.S. Marine Corps
Director, MAGTF Warfighting Center
Marine Corps Combat Development Command
Quantico, Virginia

DISTRIBUTION: 139 000500 00

Record of Changes

Change No.	Date of Change	Date of Entry	Organization	Signature

(reverse blank)

Marine Rifle Squad

Table of Contents

Chapter 1. Organization and Armament

Paragraph		Page
1001	General	1-1
1002	Organization of the Rifle Squad	1-1
1003	Composition of the Rifle Squad	1-1
1004	Weapons	1-1
1005	Duties of Individuals	1-4
1006	Fire Support for the Squad	1-6

Chapter 2. Technique of Fire

Section I. Introduction

2101	General	2-1
2102	Training	2-1

Section II. Range Determination

2201	Importance and Methods	2-2
2202	Estimation by Eye	2-2
2203	Five-Degree Method	2-4
2204	Observation of Fire	2-4

Section III. Rifle and Automatic Rifle Fire and Its Effect

2301	General	2-6
2302	Trajectory	2-6
2303	Cone of Fire	2-7
2304	Beaten Zone	2-8
2305	Classes of Fire	2-8
2306	Effect of Rifle Fire	2-11
2307	Rates of Fire	2-11

Section IV. M-203 Grenade Launcher

2401	General	2-13
2402	Employment	2-13
2403	Trajectory	2-14
2404	Firing Positions	2-14
2405	Methods of Firing	2-15
2406	Effect of Grenade Launcher Fire	2-16

Section V. Fire Commands

2501	Purpose and Importance	2-17
2502	Elements	2-17
2503	Alert	2-18
2504	Direction	2-18
2505	Target Description	2-21
2506	Range	2-21
2507	Target Assignment	2-21
2508	Fire Control	2-23
2509	Signals	2-23
2510	Delivery of Fire Commands	2-23
2511	Subsequent Fire Commands	2-24

Section VI. Application of Fire

2601	General	2-26
2602	Types of Unit Fire	2-26
2603	Fire Delivery	2-30
2604	Reduced Visibility Firing	2-33
2605	Rate of Fire	2-34
2606	Fire Control and Fire Discipline	2-35

Chapter 3. Combat Formations and Signals

Section I. Combat Formations

3101	General	3-1
3102	Basic Combat Formations	3-1
3103	Changing Formations (Battle Drill)	3-12

Section II. Signals

3201	General	3-35
3202	Whistle	3-35
3203	Special	3-35
3204	Arm-and-Hand	3-35

Chapter 4. Offensive Combat

Section I. General

4101	Purpose	4-1
4102	Phases of Offensive Combat	4-1

Section II. Preparation Phase

4201	General	4-2
4202	Movement to the Assembly Area	4-2
4203	Final Preparations in the Assembly Area	4-12
4204	Movement to the Line of Departure	4-16
4205	Special Situation	4-20

Section III. Conduct Phase

4301	General	4-21
4302	Movement From the Line of Departure to the Assault Position	4-21
4303	Movement from the Assault Position Through the Objective	4-25
4304	Enemy Counterattack	4-29
4305	Consolidation	4-30
4306	Reorganization	4-32

Section IV. Exploitation Phase

4401	Exploitation	4-33

Section V. Night Attack

4501	General	4-34
4502	Tactical Control Measures	4-34
4503	Security Patrols	4-36

4504	Preparatory Phase for the Night Attack	4-36
4505	Conduct Phase of the Night Attack	4-37
4506	Consolidation and Reorganization Phase of the Night Attack	4-38

Section VI. Infiltration

4601	General	4-39
4602	Planning and Preparation	4-39
4603	Conduct of the Attack by Infiltration	4-42

Chapter 5. Defensive Combat

Section I. General

5101	Purpose	5-1
5102	Mission	5-1
5103	Definitions	5-2
5104	Fundamentals of Defense	5-7
5105	Defensive Missions of the Squad	5-8
5106	The Fire Team in the Defense	5-9
5107	The Rifle Squad in the Defense	5-14

Section II. Defensive Procedures

5201	Troop Leading Procedures in the Defense	5-18
5202	Squad Plan of Defense	5-18
5203	Squad Security	5-20
5204	Organization of the Ground	5-20
5205	Squad Defense Order	5-28
5206	Conduct of the Defense	5-30
5207	Defense Against Mechanized Attack	5-32
5208	Movement to Supplementary Fighting Positions	5-32
5209	Local Security for Platoons and Companies	5-33
5210	Security Forces	5-33

Chapter 6. Amphibious Operations

6001	Introduction	6-1
6002	Preembarkation	6-1
6003	Duties Aboard Ship	6-1
6004	Debarkation	6-3
6005	Movement From Ship To Shore	6-7
6006	Amphibious Assault	6-8

Chapter 7. Helicopterborne Operations

7001	Introduction	7-1
7002	Concept of Employment	7-1
7003	Basic Definitions	7-1
7004	Helicopterborne Operations Training	7-2
7005	Conduct of the Helicopterborne Assault	7-5

Chapter 8. Patrolling

Section I. Patrol Organization

8101	General	8-1
8102	General Organization	8-1
8103	Special Organization	8-2
8104	Task Organization	8-3

Section II. Patrol Preparations

8201	Mission	8-4
8202	Platoon Commander's Responsibilities	8-4
8203	Patrol Leader's Preparation	8-6
8204	Study the Mission	8-6
8205	Plan Use of Time	8-7
8206	Study the Terrain and Situation	8-8
8207	Organize the Patrol	8-8
8208	Select Men, Weapons, and Equipment	8-8
8209	Issue the Warning Order	8-9
8210	Coordinate (Continuous Throughout)	8-11
8211	Make Reconnaissance	8-12
8212	Complete Detailed Plans	8-12
8213	Issue Patrol Order	8-14
8214	Supervise (Continuous), Inspect, Rehearse, and Reinspect	8-16

Section III. Conduct of Patrols

8301	Formation and Order of Movement	8-18
8302	Departure and Reentry to Friendly Lines/Areas	8-18
8303	Exercise of Control	8-20
8304	Navigation	8-21

8305	Security	8-21
8306	Movement Control Measures	8-23
8307	Actions at Danger Areas	8-25
8308	Immediate Actions Upon Contact With the Enemy	8-25

Section IV. Reconnaissance Patrols

8401	General	8-30
8402	Missions	8-30
8403	Types of Reconnaissance	8-30
8404	Task Organization of Reconnaissance Patrols	8-31
8405	Equipment	8-31
8406	Actions at the Objective	8-32

Section V. Combat Patrols

8501	General	8-34
8502	Types of Combat Patrols and Their Missions	8-34
8503	Task Organization of Combat Patrols	8-34
8504	Equipment	8-35
8505	Contact Patrols	8-35
8506	Ambush Patrols	8-36
8507	Security Patrols	8-41

Section VI. Information and Reports

8601	General	8-43
8602	Sending Information	8-43
8603	Captured Documents	8-44
8604	Patrol Report	8-44
8605	Patrol Critique	8-44

Chapter 9. Special Tactics and Techniques

Section I. Military Operations on Urbanized Terrain

9101	General	9-1
9102	Structural Classification	9-2
9103	Tactical Considerations	9-4
9104	Phases of Attacking a Built-Up Area	9-6
9105	Organization of the Rifle Squad	9-6

9106	Search Party Procedures	9-7
9107	Techniques Employed	9-9
9108	How to Prepare a Building for Defense	9-12

Section II. Attack of Fortified Areas

| 9201 | General | 9-15 |
| 9202 | Squad Assigned Mission of Seizing an Objective | 9-20 |

Section III. Tank-Infantry Coordination

9301	General	9-23
9302	Tank Capabilities and Limitations	9-23
9303	Tank and Infantry Teamwork	9-24

Section IV. Mine Warfare and Demolitions

| 9401 | Mine Warfare | 9-28 |
| 9402 | Demolitions | 9-31 |

Section V. Nuclear and Chemical Defense

| 9501 | Introduction to Nuclear Defense | 9-34 |
| 9502 | Chemical Defense | 9-38 |

Section VI. Guerilla Operations

9601	General	9-43
9602	Characteristics	9-43
9603	Indoctrination for the Squad	9-43
9604	Establishing a Patrol Base	9-44
9605	Establishment of Control Over Civil Populace	9-47
9606	Patrol Operations Against Guerrillas	9-48
9607	Offensive Action Against Guerrillas	9-48
9608	Attacking Guerrilla Houses	9-49

Appendixes:

A	Military Symbols	A-1
B	Method of Challenging by Sentries	B-1
C	Troop Leading Procedures	C-1
D	Estimate of the Situation	D-1
E	Squad Five-Paragraph Order	E-1
F	Reporting Information	F-1
G	Handling Prisoners of War	G-1
H	Cover, Concealment, and Camouflage	H-1
I	Glossary	I-1
J	References	J-1

Index

Index-1

Chapter 1

Organization and Armament

1001. General

The mission of the rifle squad is to locate, close with, and destroy the enemy by fire and maneuver, or repel the enemy's assault by fire and close combat.

1002. Organization of the Rifle Squad

The rifle squad consists of three fire teams, each of which is built around an automatic weapon and controlled by a fire team leader.

1003. Composition of the Rifle Squad

The squad is composed of 13 men: a sergeant (squad leader) and three fire teams of four men each. Each fire team consists of a corporal (fire team leader/grenadier), two lance corporals (automatic rifleman and assistant automatic rifleman), and a private or private first class (rifleman). (See fig. 1-1.)

1004. Weapons

    a. Organic Weapons. The organic weapons of the squad are as follows:

        (1) Squad Leader. M-16 rifle and bayonet knife.

        (2) Fire Team Leader/Grenadier. M-16 rifle with a 40 mm, M-203 grenade launcher attached and bayonet knife.

        (3) Automatic Rifleman. Squad automatic weapon (SAW) and combat knife (K-bar).

        (4) Assistant Automatic Rifleman. M-16 rifle and bayonet knife.

        (5) Rifleman. M-16 rifle and bayonet knife.

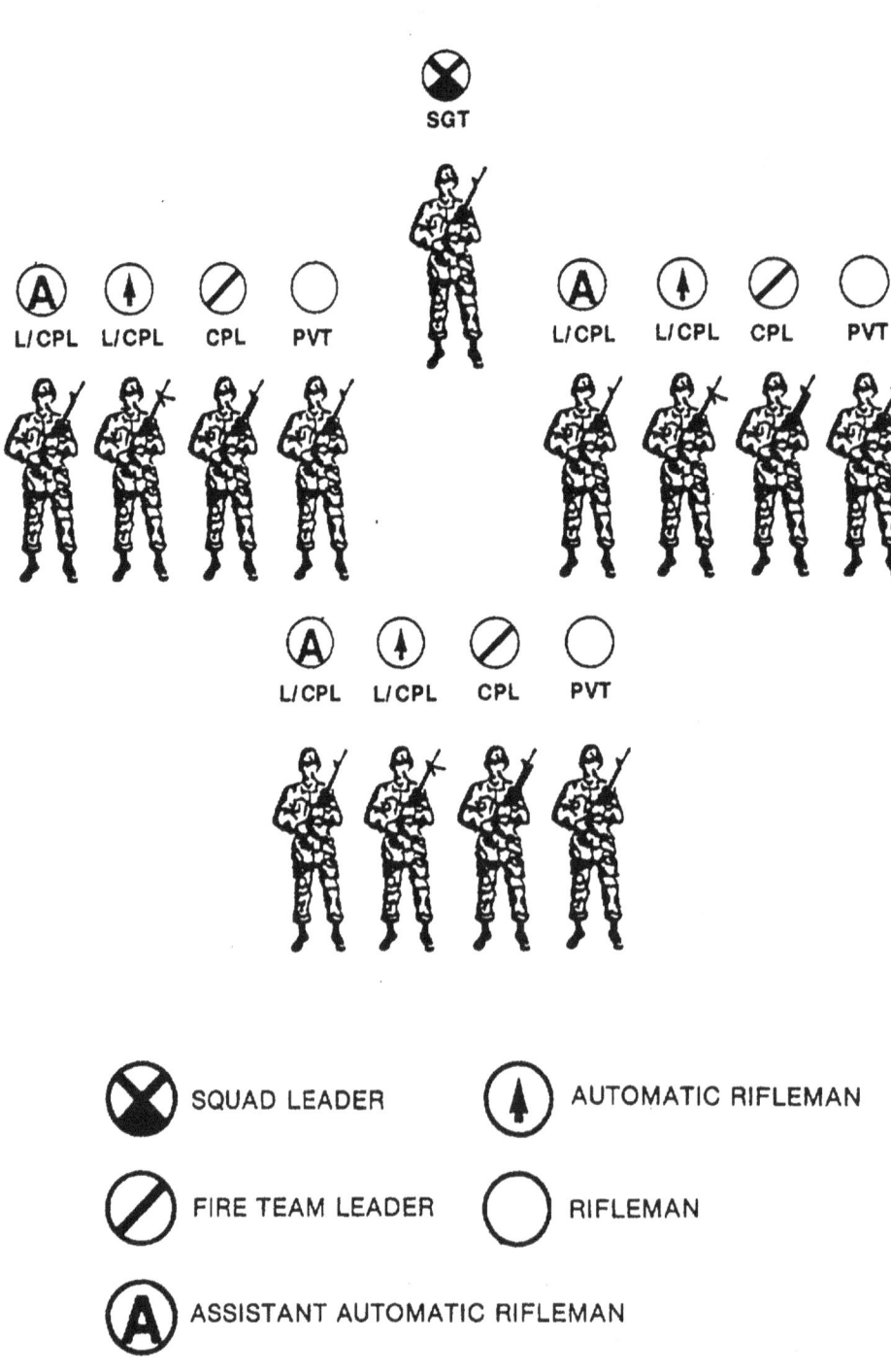

Figure 1-1. Marine Rifle Squad.

b. Ammunition

(1) M-16 Rifle

(a) Cartridge, 5.56 mm, ball, M-855.

(b) Cartridge, 5.56 mm, tracer, M-856.

(c) Cartridge, 5.56 mm, blank, M-200.

(d) Cartridge, 5.56 mm, ball, M-193 (to be used until existing stocks are depleted).

(e) Cartridge, 5.56 mm, tracer, M-196 (to be used until existing stocks are depleted).

(2) Squad Automatic Weapon

(a) Cartridge, 5.56 mm, linked, 4 and 1, M-855/M-856.

(b) Cartridge, 5.56 mm, blank, linked, M-200 with M-27 links.

(c) Unlinked M-193/M-196 (ball/tracer) cartridges may be fired from 30-round magazines, but reliability is greatly reduced. Firing should be limited to emergency situations.

(3) M-203 Grenade Launcher

(a) Cartridge, 40 mm, high explosive dual purpose (HEDP), M-433E1 (antipersonnel/antiarmor).

(b) Cartridge, 40 mm, training practice, M-407A1.

(c) Riot control and signaling cartridges.

(d) Cartridge, 40 mm, high explosive (HE), M-406 (to be used until existing stocks are depleted).

(e) Cartridge, 40 mm, HE airburst, M-397 (to be used until existing stocks are depleted).

(f) Cartridge, 40 mm, multiple projectile, M-576 (antipersonnel) (to be used until existing stocks are depleted).

(g) Cartridge, 40 mm, smokeless and flackless, M-463 (antiarmor) (to be used until existing stocks are depleted).

c. Supplementary Weapons and Munitions. The following is a list of weapons and munitions available to members of the squad:

- Demolitions.
- Claymore mines.
- Hand grenades (fragmentation, smoke, and gas [CS]).
- Light assault weapons.
- Ground signals and flares.

d. The following weapons can be employed in conjunction with the squad:

- 7.62 mm machine gun, (M-60).
- Shoulder-launched, multipurpose assault weapon (SMAW), 83 mm assault rocket launcher.
- .50 caliber machine gun (M-2).
- 40 mm machine gun (MK-19).
- 60 mm/81 mm mortars.
- Dragon/TOW antitank weapons.
- Artillery.
- Naval gunfire.
- Close air support from fixed-wing aircraft and close-in fire support from attack helicopters.

1005. Duties of Individuals

a. General. Every man of a fire team must know the duties of the other team members, and in turn, the fire team leader and the squad leader should be able to assume the duties of their next superior.

b. Squad Leader. The squad leader carries out the orders issued to him by the platoon commander. He is responsible for the discipline, appearance, training, control, conduct, and welfare of his squad at all times, as well as the condition, care, and economical use of its weapons and equipment. In combat, he is also responsible for the tactical employment, fire discipline, fire control, and maneuver of his squad. He takes position where he can best carry out the orders of the platoon commander and observe and control the squad.

c. Fire Team Leader/Grenadier. The fire team leader carries out the orders of the squad leader. He is responsible for the fire discipline and control of his fire team and for the condition, care, and economical use of its weapons and equipment. In carrying out the orders of the squad leader, he takes a position to best observe and control the fire team. Normally, he is close enough to the automatic rifleman to exercise effective control of his fires. In addition to his primary duties as a leader, but not to the detriment of them, he serves as a grenadier and is responsible for the effective employment of the grenade launcher, his rifle, and for the condition and care of his weapons and equipment. The senior fire team leader in the squad serves as assistant squad leader.

d. Automatic Rifleman. The automatic rifleman carries out the orders of the fire team leader. He is responsible for the effective employment of the automatic rifle and for the condition and care of his weapon and equipment.

e. Assistant Automatic Rifleman. The assistant automatic rifleman assists in the employment of the automatic rifle. He carries additional magazines and/or ammunition boxes for the automatic rifle and is prepared to assume the duties of the automatic rifleman. He is responsible for the effective employment of his rifle and for the condition and care of his weapon and equipment.

f. Rifleman. The rifleman in the fire team carries out the orders of the fire team leader. He is responsible for the effective employment of his rifle and for the condition and care of his weapon and equipment. The rifleman is trained as a scout.

b. Squad Leader. The squad leader carries out the orders issued to him by the platoon commander. He is responsible for the discipline, appearance, training, control, conduct, and welfare of his squad at all times, as well as the condition, care, and economical use of its weapons and equipment. In combat, he is also responsible for the tactical employment, fire discipline, fire control, and maneuver of his squad. He takes position where he can best carry out the orders of the platoon commander and observe and control the squad.

c. Fire Team Leader/Grenadier. The fire team leader carries out the orders of the squad leader. He is responsible for the fire discipline and control of his fire team and for the condition, care, and economical use of its weapons and equipment. In carrying out the orders of the squad leader, he takes a position to best observe and control the fire team. Normally, he is close enough to the automatic rifleman to exercise effective control of his fires. In addition to his primary duties as a leader, but not to the detriment of them, he serves as a grenadier and is responsible for the effective employment of the grenade launcher, his rifle, and for the condition and care of his weapons and equipment. The senior fire team leader in the squad serves as assistant squad leader.

d. Automatic Rifleman. The automatic rifleman carries out the orders of the fire team leader. He is responsible for the effective employment of the automatic rifle and for the condition and care of his weapon and equipment.

e. Assistant Automatic Rifleman. The assistant automatic rifleman assists in the employment of the automatic rifle. He carries additional magazines and/or ammunition boxes for the automatic rifle and is prepared to assume the duties of the automatic rifleman. He is responsible for the effective employment of his rifle and for the condition and care of his weapon and equipment.

f. Rifleman. The rifleman in the fire team carries out the orders of the fire team leader. He is responsible for the effective employment of his rifle and for the condition and care of his weapon and equipment. The rifleman is trained as a scout.

Chapter 2

Technique of Fire

Section I. Introduction

2101. General

When squad members have completed individual marksmanship training, and before they commence combat firing in tactical problems, they must learn the techniques of rifle, automatic rifle, and grenade launcher fire. Technique of fire refers to the application and control of the combined fire of a fire unit. A fire unit is a group of men whose combined fire is under the direct and effective control of a leader. The fire units discussed in this manual are the Marine rifle squad and its fire teams.

2102. Training

The steps in technique of fire training and the sequence in which they are given are listed below. Detailed information can be found in FMFM 0-8, *Basic Marksmanship*, (under development) and FMFM 1-3A, *Field Firing Techniques*, (to be superseded by FMFM 0-9).

   a. Target detection.

   b. Field firing positions.

   c. Range estimation.

   d. Rifle and automatic rifle fire and its effect.

   e. Grenade launcher fire and its effect.

   f. Fire commands.

   g. Application of fire.

   h. Field target firing.

Section II. Range Determination

2201. Importance and Methods

Range determination is a process of finding out the approximate distance from an observer to a target or any distant object. Accurate range determination allows the squad members to set their sights correctly and place effective fire on enemy targets. Three methods of determining ranges are estimation by eye, the five-degree method, and observation of fire.

2202. Estimation by Eye

There are two methods used in estimating range by eye: the mental unit of measure and the appearance of objects. With training and practice, accurate ranges can be determined and a high volume of surprise fire can be delivered on the enemy.

a. Mental Unit of Measure

(1) To use the mental unit of measure, the Marine visualizes a 100-meter distance, or any other unit of measure familiar to him. With this unit in mind, he mentally determines how many of these units there are between his position and the target. (See fig. 2-1.) In training, mental estimates should be checked by pacing off the distance. The average man takes about 130 steps per 100 meters.

(2) Distances beyond 500 meters can most accurately be estimated by selecting a halfway point, estimating the range to this halfway point, then doubling it. (See fig. 2-1.)

b. Appearance of Objects. When there are hills, woods, or other obstacles between the observer and target or where most of the ground is hidden from view, it is impractical to apply the mental unit of measure to determine range. In such cases, another method, based on appearance of objects, may be used. Through practice, the Marine learns how objects familiar to him appear at various known ranges. For example, watch a man when he is standing 100 meters away. Fix the appearance of his size and the details of his features and equipment firmly in mind. Watch him in the kneeling position, then in the prone position. By comparing the appearance of a man at 100, 200, 300, and 500 meters, a series of mental pictures is established. When time and conditions permit, accuracy can be improved by averaging a number of estimates by different men to determine the range.

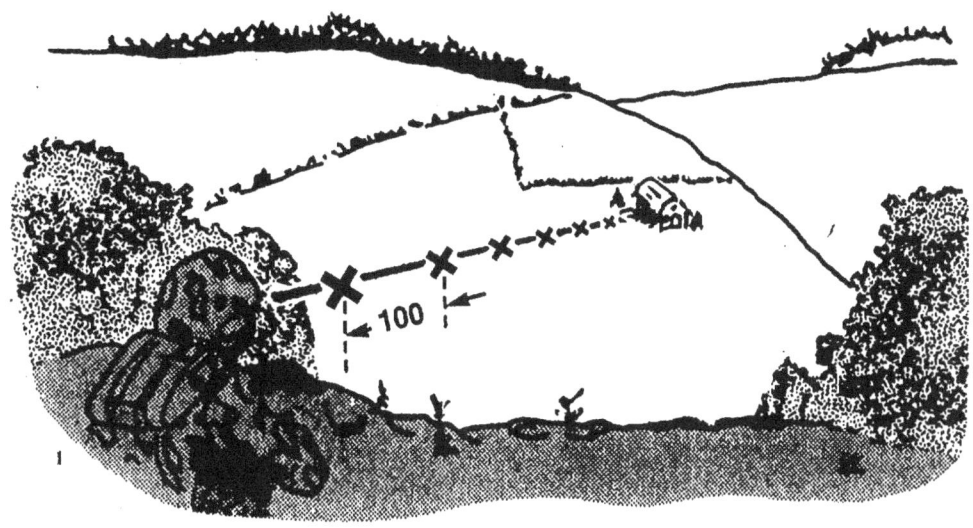

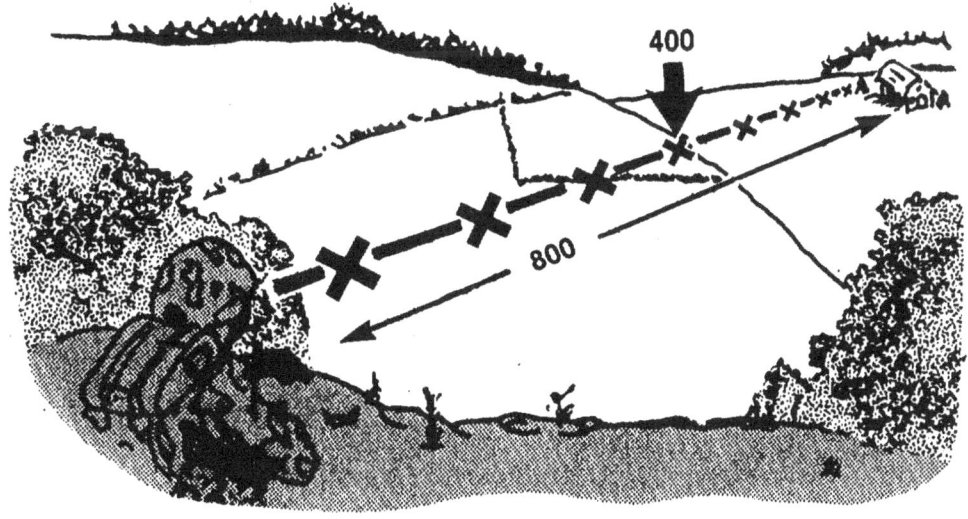

Figure 2-1. Estimation by Eye.

2203. Five-Degree Method

If time and tactical conditions permit, a relatively accurate method of determining range is the five-degree method. All that is required is a lensatic compass and the knowledge of how to pace off distances.

a. Select and mark a start point.

b. From the start point, shoot an azimuth to the object (target point) to which you are trying to determine the range. Note the azimuth.

c. Standing at the start point, turn right (or left) until the reading on the compass is 90 degrees greater (or less) than the azimuth to the target point.

d. Walk at a right angle to the line between the start point and the target point; stop periodically to shoot an azimuth to the target point.

e. When the compass shows a difference of five degrees from the original start point to target point azimuth, turn back toward the start point. It is critical that the reading of the compass be precise.

f. Walk back toward the start point using a 36- to 40-inch pace (approximately one meter); count the number of paces to the start point.

g. When you reach the start point, multiply the number of paces you counted by eleven. For example, if you counted 100 paces, multiply 100 by 11. The range to the object is approximately 1100 meters. (See fig. 2-2.)

2204. Observation of Fire

Accurate range determination can be made by observing the strike of tracer or ball ammunition. An observer is necessary because it is difficult for a rifleman to follow his own tracer and see its impact. If this method is used, the range may be estimated quickly and accurately, but the possibility of achieving surprise is lost and the firing position may be revealed to the enemy. This procedure requires that—

a. The Marine firing estimates the range by eye, sets his sight for that range, and fires.

b. The observer follows the path of the tracer and notes the impact of the round.

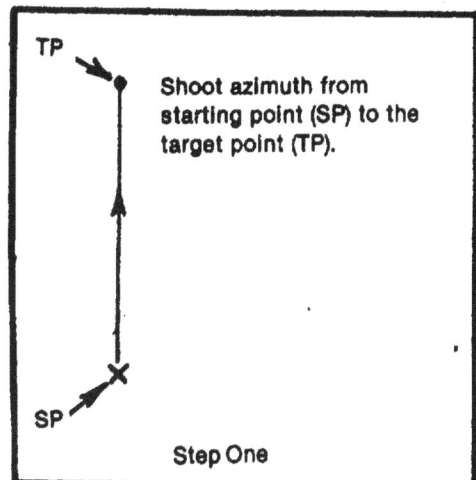
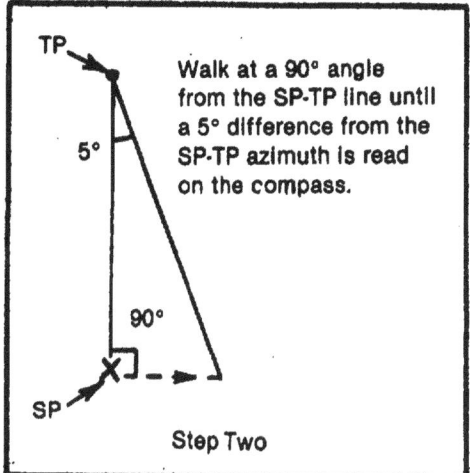
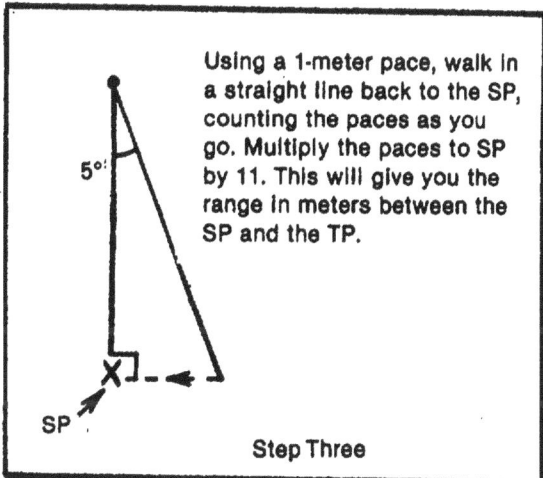

Figure 2-2. Five-Degree Method Range Determination.

c. The observer calls out sight corrections in clicks of elevation and windage necessary to hit the target.

d. The Marine firing makes sight corrections, and continues to fire and make corrections until a hit on the target is observed. The observer keeps track of the number of clicks of elevation made in getting the round onto the target.

e. The final sight setting to hit the target (with consideration to the zero of the rifle) indicates the range to the target. The Marine firing announces the range by voice or signal.

Section III. Rifle and Automatic Rifle Fire and Its Effect

2301. General

Rifle and automatic rifle fire and its effect comprise the second step in technique of fire training. A knowledge of what the bullet does while it is in flight and an understanding of the effects of fire on the enemy can assist the rifleman or automatic rifleman in obtaining maximum effectiveness.

2302. Trajectory

Trajectory is the path of a bullet in its flight through the air. The trajectory is almost flat at short ranges, but as range increases, the height of the trajectory increases. (See fig. 2-3.)

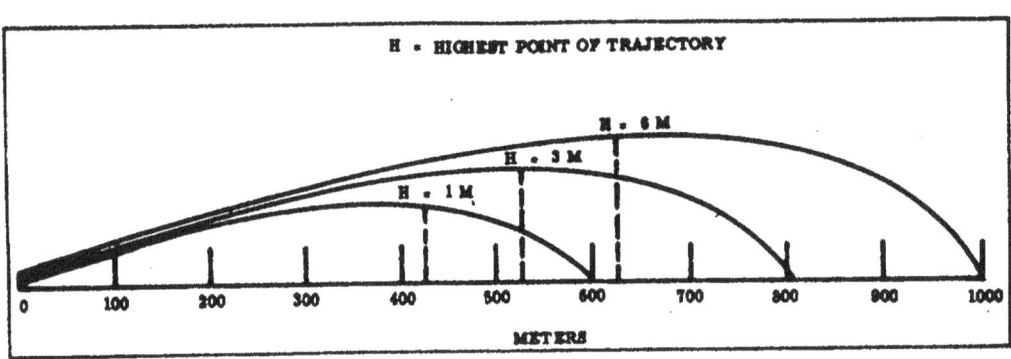

Figure 2-3. Diagram Showing Maximum Ordinate (H) of Trajectory.

The space between the rifle and the target, in which the trajectory does not rise above the height of an average man (68 inches), is called the danger space. (See fig. 2-4.) A bullet fired from the rifle at ground level (prone) at a target located at relatively short range gives continuous danger space, providing the ground is level or slopes uniformly. At greater ranges, only parts of the space between the rifle and the target are danger space because the trajectory of the bullet rises above the head of a man of average height. When the trajectory of the bullet does rise above the head of an average man, this is called dead space.

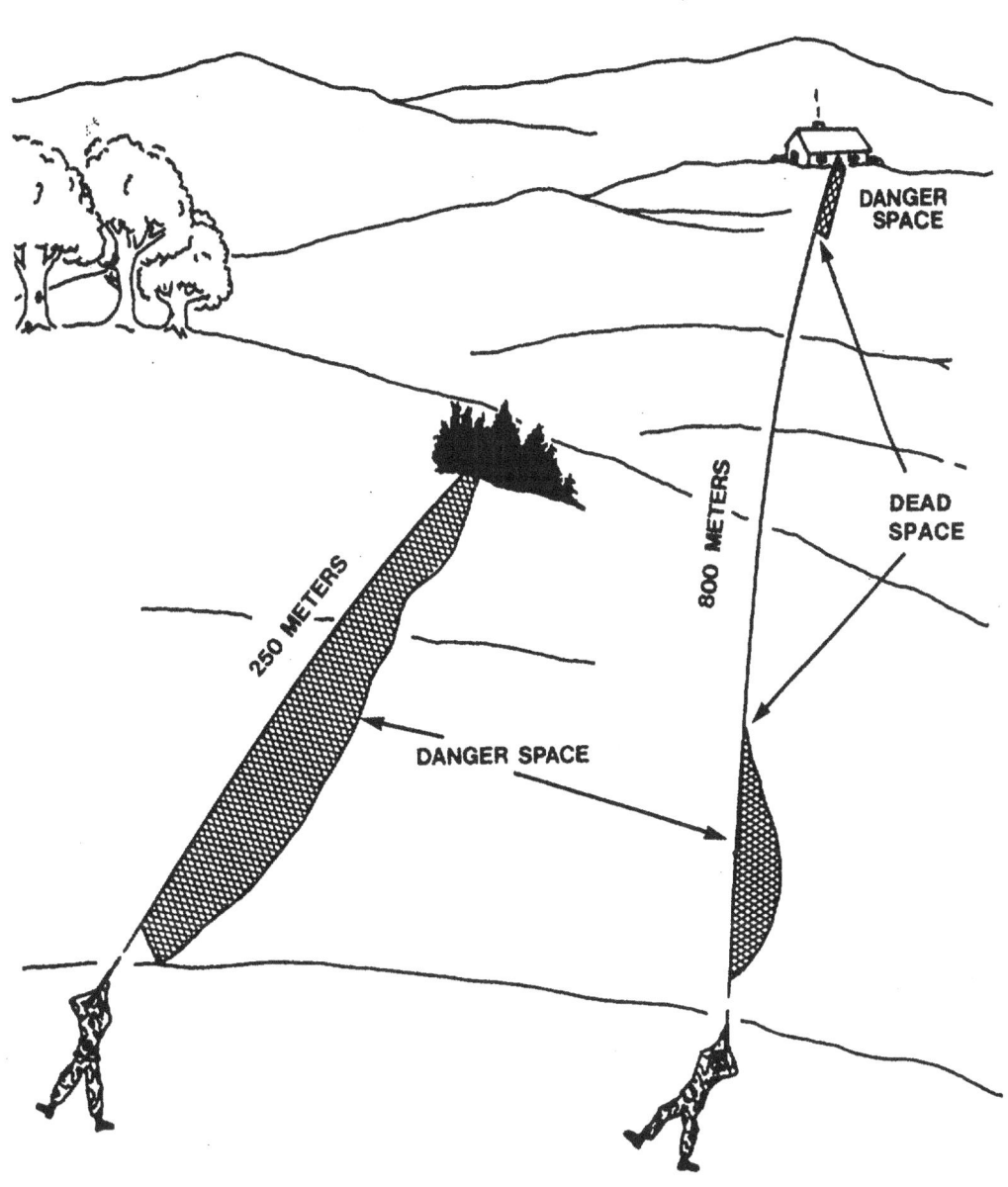

Figure 2-4. Danger Space/Dead Space.

2303. Cone of Fire

Each bullet fired from a rifle at the same target follows a slightly different path or trajectory through the air. The small differences in trajectories are caused by slight variations in aiming, holding, trigger squeeze, powder charge, wind, or atmosphere. As the bullets leave the muzzle of a weapon, their trajectories form a cone shaped pattern known as the cone of fire. (See fig. 2-5.)

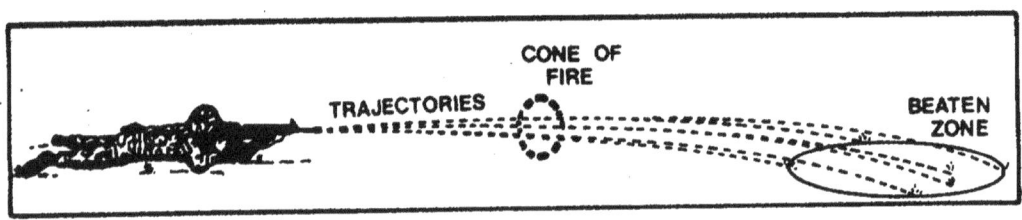

Figure 2-5. Cone of Fire and Beaten Zone.

2304. Beaten Zone

The cone of fire striking a horizontal target forms a beaten zone which is long and narrow in shape. Beaten zones on horizontal targets vary in length. As range increases, the length of the beaten zone decreases. The slope of the ground affects the size and shape of the beaten zone. Rising ground shortens the beaten zone; ground sloping downward at an angle less than the curve of the trajectories lengthens the beaten zone. Ground that falls off at an angle greater than the fall of the bullets will not be hit and is said to be in defilade.

2305. Classes of Fire

Rifle fire is classified both with respect to the target (direction) and with respect to the ground.

    a. Fire With Respect to the Target. (See figs. 2-6 and 2-7.)

        (1) **Frontal Fires.** Fires delivered perpendicular to the front of a target.

        (2) **Flanking Fires.** Fires delivered against the flank of a target.

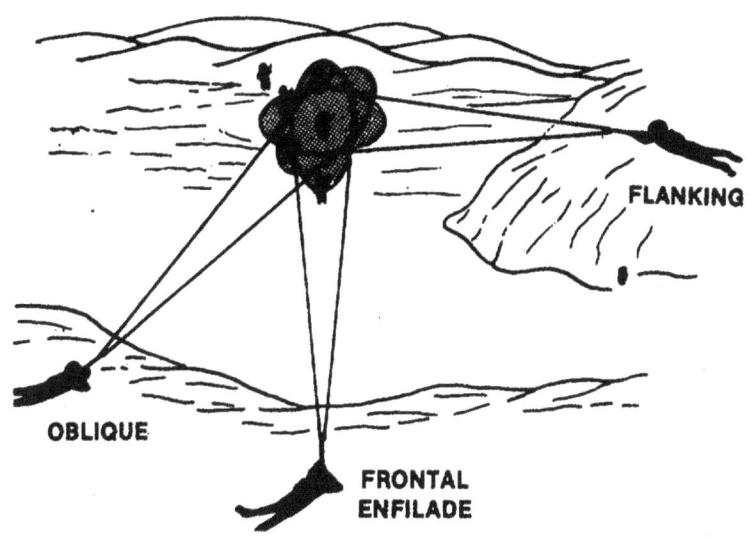

Figure 2-6. Fire With Respect to the Target.
(Example 1)

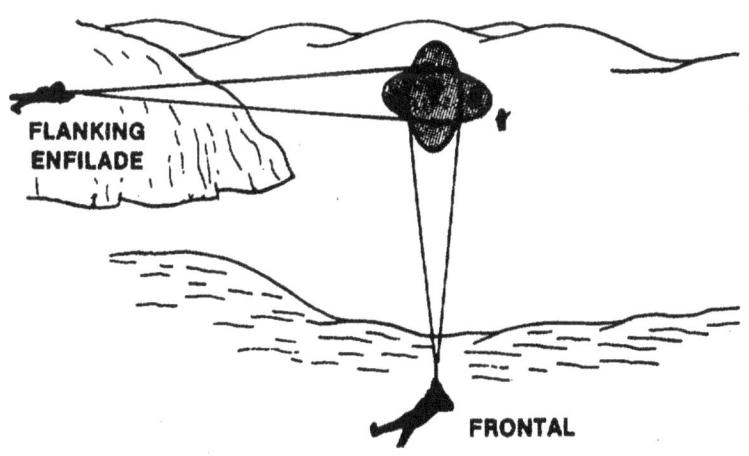

Figure 2-7. Fire With Respect to the Target.
(Example 2)

(3) Enfilade Fire. Fire delivered so that the long axis of the beaten zone coincides or nearly coincides with the long axis of the target. Enfilade fire may be either flanking or frontal.

b. Fire With Respect to the Ground. (See fig. 2-8.)

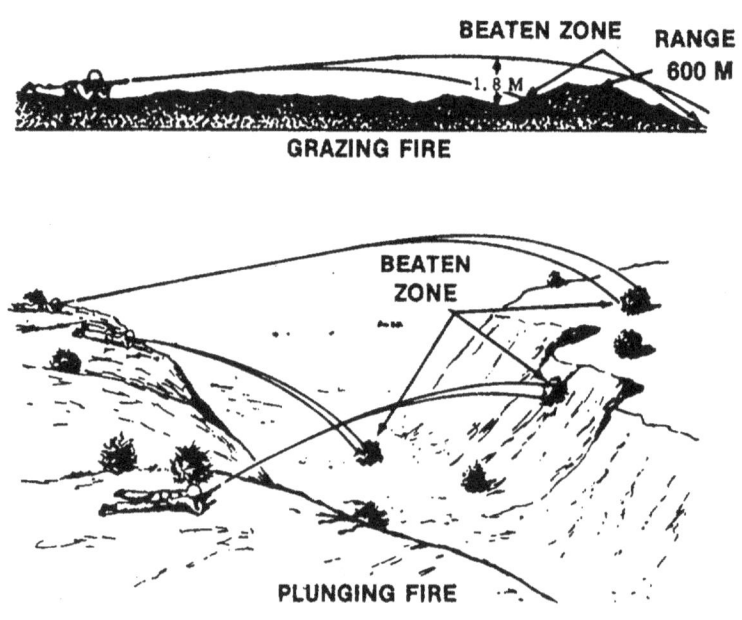

Figure 2-8. Fire With Respect to the Ground.

(I) Grazing Fires. Grazing fires do not rise above the height of a standing man. Rifle fire from the prone position may provide grazing fire at ranges up to 600 meters over level or uniformly sloping ground.

(2) Plunging Fires. Plunging fires strike the ground at a high angle so that the danger space is practically confined to the beaten zone and the length of the beaten zone is shortened. Fire at longer ranges becomes increasingly plunging because the angle of fall of the bullets becomes greater. Fire from high ground to a target on low ground may be plunging fire. Fire into abruptly rising ground causes plunging fire at the point of impact.

(3) Overhead Fires. Overhead fires are delivered over the heads of friendly troops. Rifle fire is considered safe when the ground protects the friendly troops to the front or if they are in position at a sufficient distance below the line of fire.

2306. Effect of Rifle Fire

The best effects from rifle fire are obtained when the squad is close to the enemy. The squad should use cover and concealment offered by terrain and take advantage of the supporting fires of machine guns, mortars, and artillery to advance as near to the enemy as possible before opening fire. Normally, it should not open fire at ranges greater than 460 meters, the maximum effective range of the rifle.

Under favorable conditions, the rifle may be used against enemy groups or area targets at ranges between 460 to 1,000 meters.

The area in which the enemy is located can usually be determined by the sound of his firing. Troops may distribute continuous fire in width and depth to cover the entire area, causing the enemy to keep his head down and making his fire ineffective.

2307. Rates of Fire

The rates of fire of squad weapons combine to form the firepower of the squad. Weapons employment and squad firepower are not determined by how fast Marines can fire their weapons but how fast they can fire accurately. The squad or fire team leader must be able to control the rate and effect of his men's fire, otherwise, ammunition is wasted.

The rate of fire for weapons is expressed in rounds per minute (RPM). The following rates of fire apply to the weapons of the rifle squad.

a. Average Rate. This term refers to the average rate of aimed fire a Marine can deliver with a semiautomatic rifle or with an M-203 grenade launcher. The following average rates apply to squad weapons:

- M-16: 10 to 12 RPM.
- M-203: 5 to 7 RPM.

b. Sustained Rate. This term applies to automatic rifles and machine guns. It is the actual rate of well-directed fire a weapon can deliver for an indefinite length of time without causing a stoppage or malfunction due to overheating. The sustained rate for the SAW is 85 RPM.

c. Rapid Rate. This term applies to automatic rifles and machine guns. It is the maximum amount of controlled fire which can be delivered on target for a short period of time (usually not more than two minutes) without causing a stoppage or malfunction due to overheating. The rapid rate for the SAW is 200 RPM.

Section IV. M-203 Grenade Launcher

2401. General

The fire team leader/grenadier carries a weapon that is both rifle and grenade launcher, and he can use either or both as the situation dictates. In order for him to best employ the M-203 portion of the weapon, he must understand the trajectory of the rounds, methods of firing, and effects of the rounds.

2402. Employment

a. Offense. The fire team leader/grenadier employs the grenade launcher in the offense to destroy groups of enemy personnel and to provide close fire support in the assault in conjunction with, and to supplement, other supporting fires.

(1) The fire team leader/grenadier personally selects targets and delivers the fires of the grenade launcher during the attack. In the last 35 meters of the assault, when the fires of the grenade launcher may endanger friendly assaulting troops on the objective, the fire team leader/grenadier employs the multiprojectile antipersonnel round. The multiprojectile round can be fired from the assault skirmish line without endangering the other assaulting Marines. The fire team leader/grenadier can fire high explosive rounds at targets which are far enough away so that the exploding HE round will not endanger the assaulting squad. (HE rounds require an arming distance of approximately 30 meters.)

(2) During the assault, the fire team leader/grenadier may employ his rifle until suitable targets appear or until he has time to reload the M-203. Suitable M-203 targets are enemy automatic rifle positions, machine gun positions, and other crew-served weapons within the fire team sector. This method of employment is used when a heavy volume of fire is needed.

b. Defense. In the defense the fire team leader/grenadier's firing position should enable him to control his fire team and deliver grenade launcher fires over the entire fire team sector of fire. Primary and supplementary positions are prepared which provide maximum cover and concealment consistent with the assigned mission. Extreme care must be taken to ensure that fields of fire are cleared of obstructions which might cause premature detonation of the projectile. As the enemy approaches the defensive position, he is subjected to an ever-increasing volume of fire. Initially, the fire team leader/grenadier should use the rifle portion of the weapon. As the enemy gets nearer to friendly positions, he should use the grenade launcher. He will fire on enemy automatic weapons and enemy troops who are in defilade. This will silence an enemy base of fire and cause enemy troops to leave covered positions so the automatic riflemen can engage them.

2403. Trajectory

The grenade launcher, at ranges up to 150 meters, has a trajectory that is relatively flat; thus, the grenade launcher is fired from the shoulder in the normal manner. As the range increases, the height of the trajectory and the time of flight of the projectile increase.

2404. Firing Positions

a. The most commonly used firing positions are the prone, kneeling, fighting hole, and standing positions. Supported positions add stability to the weapon and should be used whenever possible; however, the fire team leader/grenadier must ensure that no part of the launcher touches the support.

b. There are two methods for holding the weapon:

(1) The left hand grips the magazine of the M-16 rifle with the left index finger positioned in the trigger guard of the M-203, while the right hand grips the pistol grip.

(2) The right hand grips the magazine of the M-16 rifle with the right index finger positioned in the trigger guard of the M-203, while the left hand grasps the hand grip of the barrel assembly. (See fig. 2-9.)

Figure 2-9. Grip Method.

2405. Methods of Firing

a. Aimed Fire. At ranges up to 150 meters, the grenade launcher can be fired from the shoulder in the normal manner from all positions using the sight leaf of the quadrant sight. However, to maintain sight alignment at ranges greater than 150 meters, the following adjustments are required:

(1) Use the quadrant sight at ranges in excess of 200 meters.

(2) In the modified prone position, the position of the butt of the rifle depends on the configuration of the shooter's body, the position of the shooter's hand on the weapon, and the range to the target.

(3) In other firing positions, lower the stock to an underarm position in order to maintain sight alignment.

b. Pointing Technique. The pointing technique is used to deliver a high rate of fire on area targets. Although the sights are not used in the pointing technique, the shooter must first be proficient in sighting and aiming, using the sight leaf and quadrant sight. He uses a modified underarm firing position which enables him to use his left hand for rapid reloading.

Although the pointing technique can be used by modifying any standard firing position, it is most frequently used during the assault. (See fig. 2-10.)

Figure 2-10. Pointing Technique.

2406. Effect of Grenade Launcher Fire

The high explosive grenade has an effective casualty radius of 5 meters. The effective casualty radius is defined as the radius of a circle about the point of detonation in which it may be expected that 50 percent of exposed troops will become casualties.

Section V. Fire Commands

2501. Purpose and Importance

Since enemy troops are trained in the use of cover and concealment, targets are often indistinct or invisible, seen only for a short time, and rarely remain uncovered for long. When a target is discovered, leaders and squad members must define its location rapidly and clearly. Squad members are trained to identify the target area quickly and accurately and to place a high volume of fire on it even though no enemy personnel may be visible. A small point target like an enemy sniper might be assigned to only one or two riflemen, while a target of considerable width like an enemy skirmish line requires the combined fires of the entire squad. As an aid in designating various types of targets, all members of the squad must become familiar with the topographical terms frequently used in designating targets; e.g., crest, hill, cut, fill, ridge, bluff, ravine, crossroads, road junction, and skyline. (See fig. 2-11.) When the squad or fire team leader has made a decision to fire on a target, he gives certain instructions as to how the target is to be engaged. These instructions form the fire command. The leader directs and controls the fire of his fire unit by fire commands.

2502. Elements

A fire command contains six basic elements that are always announced or implied. Fire commands for all weapons follow a similar order and include similar elements. Only essential elements are included. The six elements (ADDRAC) of the fire command are:

Alert.

Direction.

Target Description.

Range.

Target Assignment.

Fire Control.

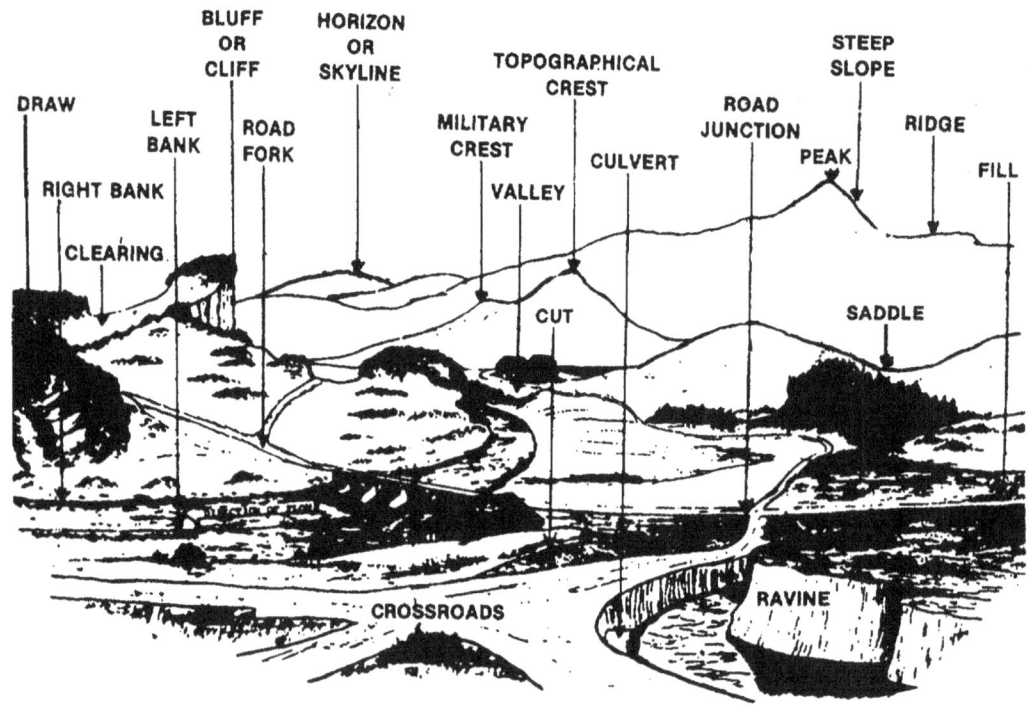

Figure 2-11. Topographical Terms.

2503. Alert

This element alerts the fire unit to be ready to receive further information. It may also tell who is to fire. Usually, it is an oral command, **SQUAD** or **FIRE TEAM**. The leader may alert only a few individuals by calling them by name. The alert may also be given by signals, personal contact, or by any other method the situation may indicate.

2504. Direction

The direction element tells which way to look to see the target. The direction of the target may be indicated in one of the following ways:

a. Orally. The general direction to the target may be given orally and should indicate the direction to the target from the unit. Figure 2-12 shows the general directions used to indicate direction orally; for example, **RIGHT FRONT**.

Figure 2-12. General Directions.

b. Tracer Ammunition

(1) Tracer ammunition is a quick and sure way to indicate direction and is the most accurate method of pinpointing targets. Whenever possible, the leader should give the general direction orally. This will direct the squad's attention to the desired area; for example:

FRONT.

WATCH MY TRACER.

(Fire 1st round) **RIGHT FLANK** (of the target).

(Fire 2d round) **LEFT FLANK** (of the target).

(2) Firing tracer ammunition to designated targets may give away the Marine's position and it will most certainly alert the enemy and reduce the advantage of surprise. To minimize the loss of surprise, the leader may wait until all other elements of the fire command are given before firing his tracer. In this case, the firing of the tracer can be the signal to commence firing.

c. Reference Points

(1) To help the members of the fire unit locate indistinct targets, the leader may use reference points to give direction to the target. He selects a reference point that is near the target and easy to recognize.

(2) When using a reference point, the word **REFERENCE** in describing the reference point and the word **TARGET** in describing the target are used. This prevents the members of the fire unit from confusing the two; for example:

SQUAD.

FRONT.

REFERENCE: ROCK PILE IN DRAW.

TARGET: SNIPER IN FIRST TREE TO THE RIGHT, ONE FIVE ZERO.

(3) When using a reference point, the direction refers to the reference point, but the range is the range to the target.

(4) Sometimes a target can best be located by using successive reference points; for example:

FIRST TEAM.

REFERENCE: STONE HOUSE. RIGHT OF STONE HOUSE, SMALL SHED.

TARGET: MACHINE GUN IN FIRST HAYSTACK RIGHT OF SHED, TWO FIVE ZERO.

d. Finger Measurement

(1) Distances across the front, known as lateral distances, are difficult to estimate in terms of meters. To measure the distance right or left of a reference point, or to measure the width of a target from one flank to another, finger measurements may be used.

(2) The method of using finger measurements is as follows:

(a) Hold the hand at arm's length directly in front of your face, palm facing away from you, index finger pointing upward.

(b) Close one eye.

(c) Select a reference point.

(d) Place one finger between the reference point and the target and then fill that space by raising more fingers until the space is covered.

(3) An example of the use of finger measurement is as follows:

SQUAD.
FRONT.
REFERENCE: TALL TREE AT EDGE OF HEDGEROW,
 RIGHT TWO FINGERS.
TARGET: MACHINE GUN, THREE HUNDRED.

2505. Target Description

The third element of the fire command is a brief and accurate description of the target.

2506. Range

Range gives the information needed to set the sight or to adjust the point of aim. The word **RANGE** is not used. Examples of range are **ONE SEVEN FIVE, TWO FIVE ZERO,** or **FOUR HUNDRED.**

2507. Target Assignment

The target assignment element tells who is to fire on the target and is broken down into two subelements as follows.

a. First, the squad leader prescribes whether the entire squad will fire on the target or whether only one or two fire teams will fire. If the unit to fire is the same as announced in the alert element, it may be omitted from the target assignment element. When the squad leader intends to alert the entire unit, but plans to use only one or two fire teams to fire on a target, the target assignment element is included.

b. The squad leader also uses this element to determine what weapons will be fired and the rate of fire for the automatic rifle. Rifles and, when fired, the M-203 always fire at the average rate. Fire team leaders normally do not fire their rifles unless it is absolutely necessary. Instead, they direct the fires of the members of their fire team on various targets within the assigned sector of fire and remain ready to transmit subsequent fire commands from the squad leader to their fire team. The following rules apply:

(1) **Automatic Riflemen.** If the squad leader wants the automatic rifles fired at the rapid rate he commands **RAPID**. If the command **RAPID** is not given, automatic rifles are fired at the sustained rate. In response to the command **RAPID**, the automatic riflemen fire initially at the rapid rate for two minutes and then change to the sustained rate. This prevents the weapon from overheating.

(2) **Fire Team Leader/Grenadier.** If the squad leader desires grenade launcher fire, he commands **GRENADIER**. If the command **GRENADIER** is not given, the fire team leaders/grenadiers do not normally fire their rifles.

c. In the following examples of the target assignment element, let us assume that in the alert element, the command **SQUAD** was given.

(1) If the target assignment element is omitted completely, all three fire teams prepare to fire as follows:

(a) Riflemen and assistant automatic riflemen fire their rifles at the average rate.

(b) Automatic riflemen fire their weapons at the sustained rate.

(2) **GRENADIER; RAPID.** All fire teams prepare to fire as follows:

(a) Riflemen, and assistant automatic riflemen prepare to fire their rifles at the average rate.

(b) Fire team leaders/grenadiers prepare to fire their M-203s at the average rate.

(c) Automatic riflemen prepare to fire their weapons at the rapid rate.

(3) **FIRST TEAM; GRENADIER; RAPID.** The first fire team prepares to fire as follows:

(a) Riflemen and assistant automatic riflemen fire their rifles at the average rate.

(b) Fire team leaders/grenadiers fire their M-203s at the average rate.

(c) Automatic riflemen fire their weapons at the rapid rate.

2508. Fire Control

The fire control element consists of a command or signal to open fire. If surprise fire is not required, the command, **COMMENCE FIRING** normally is given without a pause as the last element of the fire command. When the leader wants all his weapons to open fire at once in order to achieve maximum surprise and shock effect, he will say, **AT MY COMMAND** or **ON MY SIGNAL**. When all men are ready, the leader gives the command or signal to commence firing.

2509. Signals

Since oral commands are likely at times to be unheard because of battle noise, it is essential that the members of fire units also understand visual and other signals. These signals must be used constantly in training. Standard arm-and-hand signals applicable to fire commands are described in chapter 3.

2510. Delivery of Fire Commands

Examples of complete fire commands are as follows:

a. In this example, the squad leader wants to place a heavy volume of surprise rifle and automatic rifle (sustained rate) fire of his entire squad on an easily recognized target:

SQUAD.
FRONT.
TROOPS.
THREE HUNDRED.
AT MY SIGNAL.

b. In this example, the squad leader desires to designate the target to his entire squad, but wants only the second fire team to engage it. He desires M-203 fire on the target and the automatic rifleman to fire at the rapid rate. Because the target is indistinct, he uses a reference point.

SQUAD.
RIGHT FRONT.
REFERENCE: STONE HOUSE, RIGHT TWO FINGERS.
TARGET: MACHINE GUN TWO FIVE ZERO.
SECOND TEAM; GRENADIER; RAPID.
COMMENCE FIRING.

2511. Subsequent Fire Commands

A subsequent fire command is used by the squad leader to change an element of his initial command or to cease fire.

a. To change an element of the initial command, the squad leader gives the alert and then announces the element he desires to change. Normally, the elements that will require changing are the target assignment and/or the fire control. The following example illustrates the use of a subsequent fire command.

(1) In the following initial fire command the squad leader alerts his entire squad but only assigns one fire team to engage the target with rifle and automatic rifle (sustained rate) fire.

SQUAD.
FRONT.
TROOPS.
THREE HUNDRED.
SECOND TEAM.
COMMENCE FIRING.

(2) The squad leader now desires the entire squad to fire on the target, fire team leaders/grenadiers to fire their M-203s, and automatic riflemen to fire at the rapid rate. Note that the squad leader does not repeat **SQUAD** in the target assignment since he alerted the entire squad and wants the entire squad to fire. The squad leader's subsequent command will be as follows:

SQUAD.

GRENADIER; RAPID.

COMMENCE FIRING.

b. To have the squad cease fire, the squad leader simply commands, **CEASE FIRE.**

c. In issuing subsequent fire commands, the squad leader must keep in mind that in most cases the noise of the battlefield will prevent the squad members from hearing him. In most cases the squad leader will pass subsequent fire commands through the fire team leaders. It is for this reason that fire team leaders do not normally fire their rifles but remain attentive to the directions of the squad leader.

Section VI. Application of Fire

2601. General

The potential firepower of the 13-man squad with all members firing is conservatively estimated at 400 well-aimed rifle and automatic rifle shots or 370 well-aimed rifle and automatic rifle shots and 15 rounds from the grenade launchers per minute. The following terms are used when discussing application of fire.

a. Neutralize. To render enemy personnel incapable of interfering with a particular operation.

b. Fire Support. Fire delivered by a unit to assist or protect another unit in combat.

c. Target of Opportunity. A target which appears in combat, within range, and against which fire has not been planned.

2602. Types of Unit Fire

a. General

(1) The size and nature of a target may call for the firepower of the entire fire unit or only parts of it. The type of target suggests the type of unit fire to be employed against it. The squad leader receives his orders from the platoon leader who usually designates a specific target or targets. It is usually desirable for each squad to cover the entire platoon target to ensure adequate coverage.

(2) A fire team distributes its fire as designated by the squad leader. Normally, the squad leader orders a fire team leader to limit the fire of his team to a sector of the squad target, to engage a separate target, or to shift to a target of opportunity.

b. Concentrated Fire. Concentrated fire is fire delivered from a deployed unit at a single point target. A large volume of fire delivered at the target from different directions, causes the beaten zones of the various weapons to meet and overlap giving maximum coverage of the target. An enemy automatic weapon that has gained fire superiority over an element of a particular unit, can often be neutralized by concentrated fire from the remaining elements which are not under direct fire. (See fig. 2-13.)

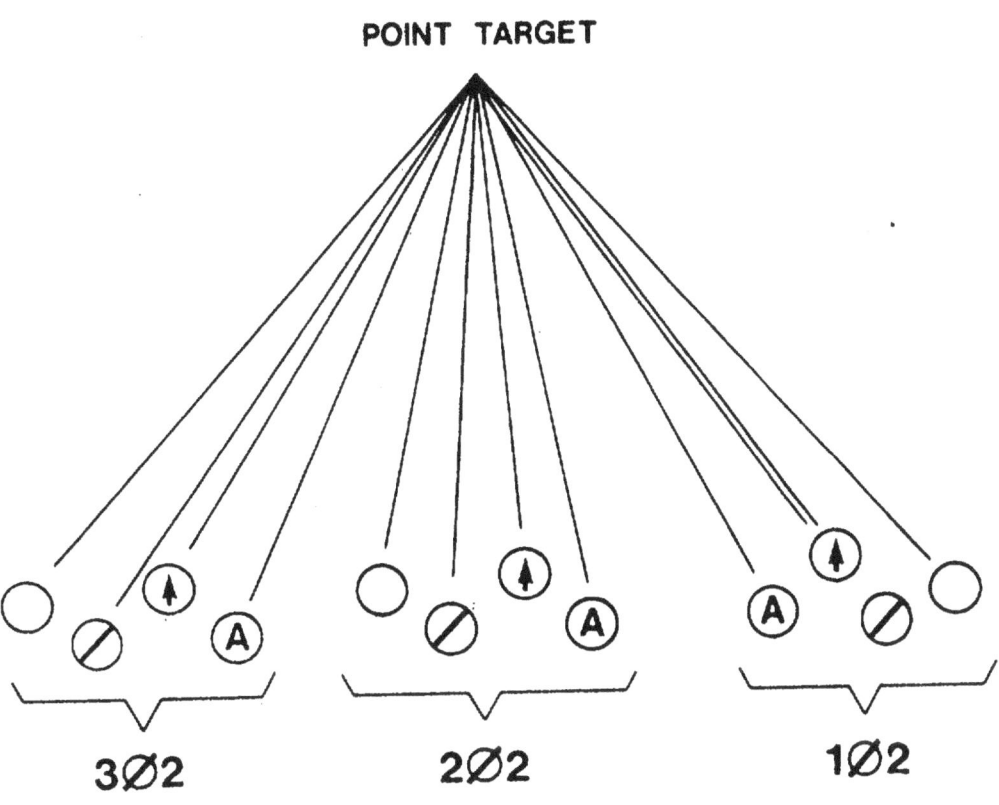

Figure 2-13. Concentrated Fire by a Rifle Squad.

c. Distributed Fire

(1) Distributed fire is fire spread in width and/or depth to keep all parts of the target under fire. Each rifleman and assistant automatic rifleman fires his first shot on that portion of the target that corresponds to his position in the squad. He then distributes his remaining shots over the remainder of the target, covering that portion of the target on which he can deliver accurate fire without changing his position. (See fig. 2-14.)

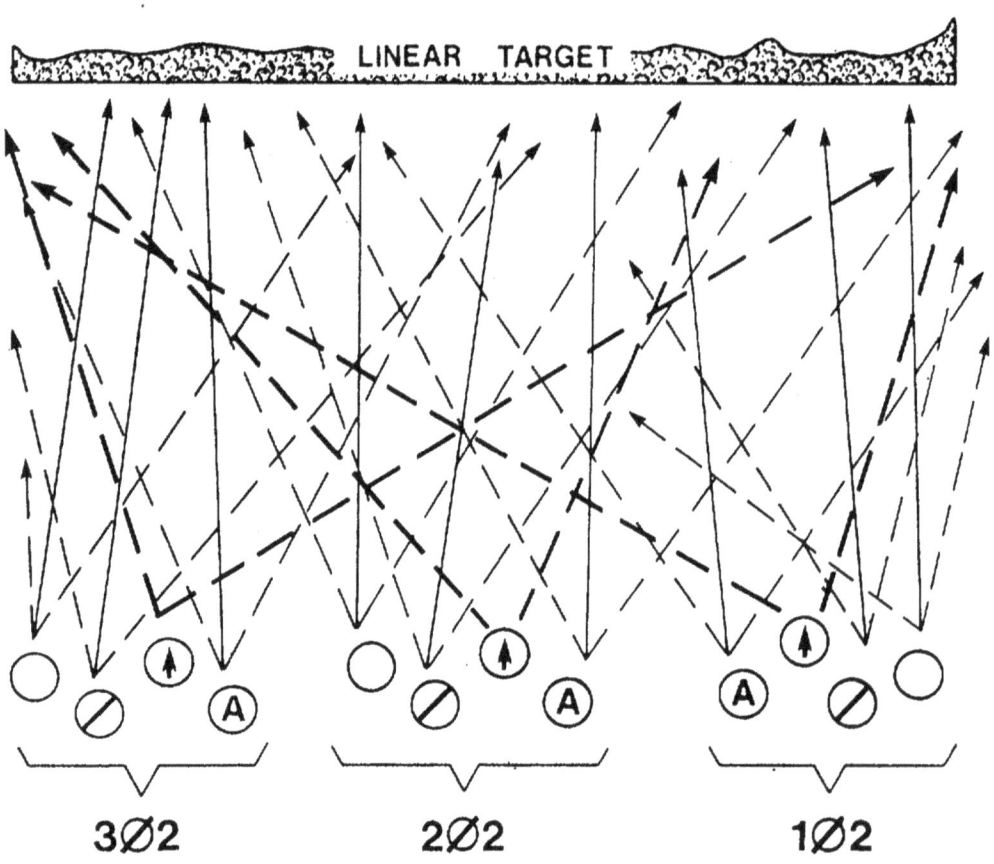

Figure 2-14. Distributed Fire by a Rifle Squad.

(2) The fire team leader/grenadier fires the first round from his grenade launcher at the center of the mass of the target. He then distributes grenades over the remaining target area.

(3) In the offense, the automatic riflemen cover the entire squad target. In the defense, the automatic riflemen cover their respective fire team's sector of fire.

(4) Distributed fire permits fire unit leaders to place the fire of their units on target so that the enemy, whether visible or not, is kept under fire. Distributed fire is the quickest and most effective method of ensuring that all parts of the target are brought under fire. (See fig. 2-15.) When

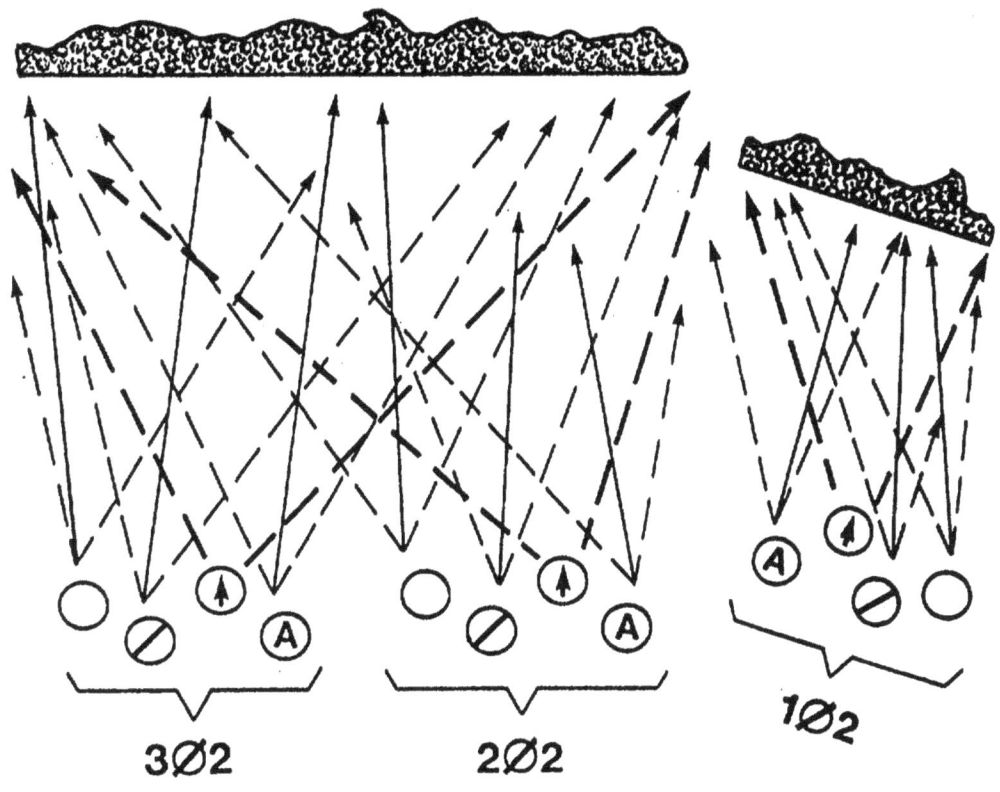

Figure 2-15. Distributed Fire by a Rifle Squad Engaging Two Separate Targets.

it becomes necessary to engage other targets, the squad leader shifts the fire of one or two fire teams as required.

d. Combinations of Concentrated and Distributed Fire. The fire team organization of the Marine rifle squad permits the squad leader to combine both concentrated and distributed fire in engaging two or more targets at the same time. As an example, the squad leader of a squad delivering distributed fire on a target could shift the fire of one or two fire teams to engage a target of opportunity with concentrated fire. (See fig. 2-16.) Whether a fire unit (squad or fire team) delivers concentrated or distributed fire is determined by the target description element of the fire command. If the target description indicates a point target (i.e., machine gun, sniper, etc.)

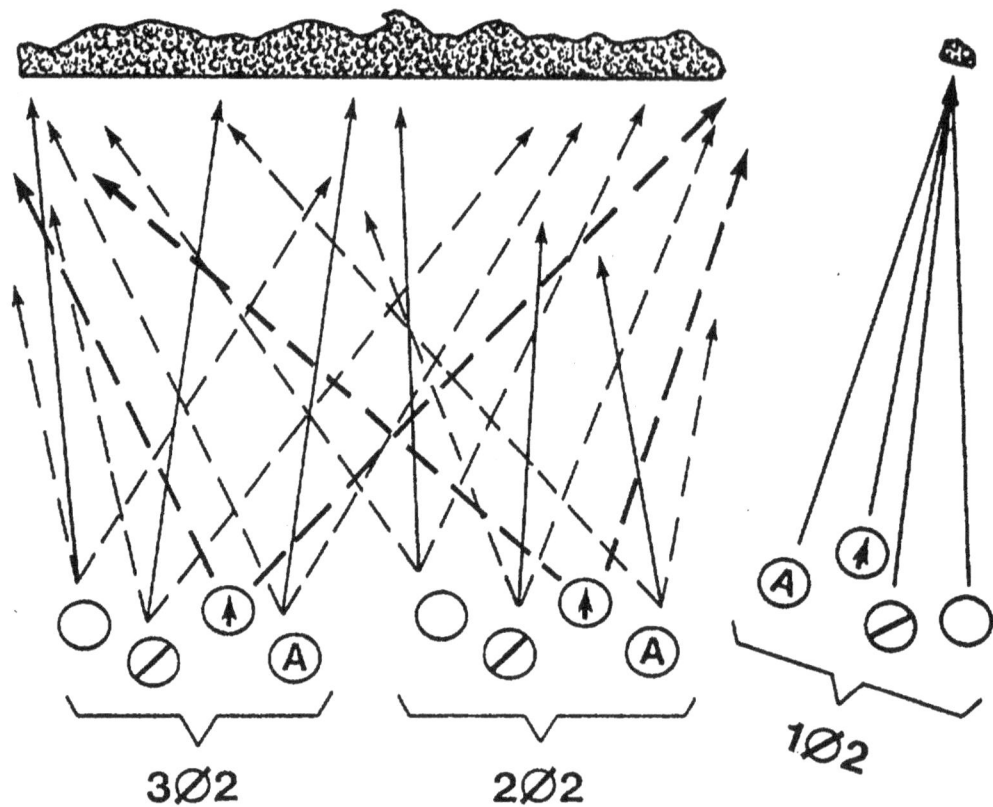

Figure 2-16. Concentrated and Distributed Fire by a Rifle Squad Engaging Two Separate Targets.

the fire unit will fire concentrated fire. If the target description indicates an area target (i.e., squad in open or dug in, or a target which the squad leader has marked the flanks), the fire unit will fire distributed fire. By assigning his fire teams fire missions using fire commands, the squad leader regulates the volume, density, and coverage of his squad's fire.

2603. Fire Delivery

a. Requirements of Position. In occupying a firing position, squads are located to satisfy the following requirements:

- Be capable of delivering desired fire support.
- Possess good fields of fire to the front.
- Have adequate cover and concealment.
- Permit fire control by the fire unit leader.

b. In the Attack

(1) Base of Fire. A base of fire covers and protects the advance of maneuvering units with its fire. Whenever possible, the fire unit that is to establish the base of fire moves undetected into a firing position. A high volume of surprise fire from an unexpected direction has a much greater psychological and physical effect than fire delivered from a known position. The leader of the unit establishing the base of fire makes every effort to select a position that allows flanking or oblique fire to be delivered into the enemy position. When the base of fire unit is in position, the following usually takes place:

(a) A heavy volume of distributed fire is placed on the enemy position to gain fire superiority.

(b) When fire superiority has been gained and the enemy is fixed in position, the rate of fire is reduced. However, fire superiority must be maintained.

(c) When the maneuver unit nears its final coordination line, the rate of fire is increased to cause the enemy to *button up* tightly, and allow the maneuver unit to move out of the assault position and initiate its assault before the enemy has time to react.

(d) When the assaulting maneuver unit reaches the final coordination line or on signal, the base of fire either ceases, shifts its fire to another target area, or leads the assault unit across the objective and then ceases or shifts.

(2) Assault Fire. Successful advance by fire and maneuver leads naturally to an assault of the target area or objective. Assault fire is that fire delivered by a unit during its assault on a hostile position.

(a) **Riflemen and Assistant Automatic Riflemen.** Both fire well-directed shots from the pointing position. They should fire the weapon using three-round bursts or they should pull the trigger each time the left foot strikes the ground. They fire at known or suspected enemy locations on the portion of the objective that corresponds with their position in the assault formation. (See fig. 2-17.)

POINTING   UNDERARM

Figure 2-17. Assault Fire Positions.

(b) **Automatic Riflemen.** The automatic riflemen fire in three- to five-round bursts from the underarm firing position. They cover the entire squad objective. Priority of fire is given to known or suspected enemy automatic weapons. (See fig. 2-17.)

(c) **Fire Team Leader.** The fire team leader's primary concern during the assault is the control of his fire team. If he is required to fire his rifle, he fires well-directed rifle fire using the pointing technique. Once a hardened or area target presents itself, the fire team leader will commence to fire the grenade launcher using the pointing technique until the target is destroyed or neutralized, or until he cannot place effective fire on the target without endangering friendly troops.

c. In the Defense. The fire team is the basic fire unit of the rifle platoon and when practical, each individual's sector of fire covers the entire fire team sector of fire. The fire team delivers fire from positions which it must hold at all costs. Members of the unit are placed where they can obtain good fields of fire and take maximum advantage of cover and concealment. (See app. H.)

(1) Riflemen, Assistant Automatic Riflemen, and Automatic Riflemen The automatic rifles provide the bulk of the squad's firepower. They must be protected and kept in operation. These Marines are assigned to cover the entire fire team sector. In addition, each automatic rifleman is assigned a principal direction of fire.

(2) Fire Team Leader. The fire team leader's primary concern in the defense is the control of his fire team. When required to fire his rifle, he will cover the entire fire team sector with a high volume of fire while the enemy is beyond the range of the M-203. Unless restrictions are placed on firing the grenade launcher, he opens fire on profitable targets as they come in range. When the final protective fires are called for, he engages the largest mass of enemy infantry in the assigned sector.

2604. Reduced Visibility Firing

a. Rifle. Under conditions of reduced visibility, the rifle can be used to deliver preplanned fire by constructing a simple rest for the weapon. When the rifle is used for this purpose, all preparations are made during daylight. In addition to sighting the rifle and erecting the rest and stakes, sights are set and fire adjusted on the target in advance. (See fig. 2-18A.)

b. Grenade Launcher. In periods of reduced visibility, the grenade launcher can also be used effectively to deliver preplanned fires by constructing a stand. When planning these fires, the squad leader should give priority to likely avenues of approach and probable enemy assault positions. All preparations are made during daylight. The weapon is emplaced and sights adjusted prior to darkness. (See fig 2-18B.)

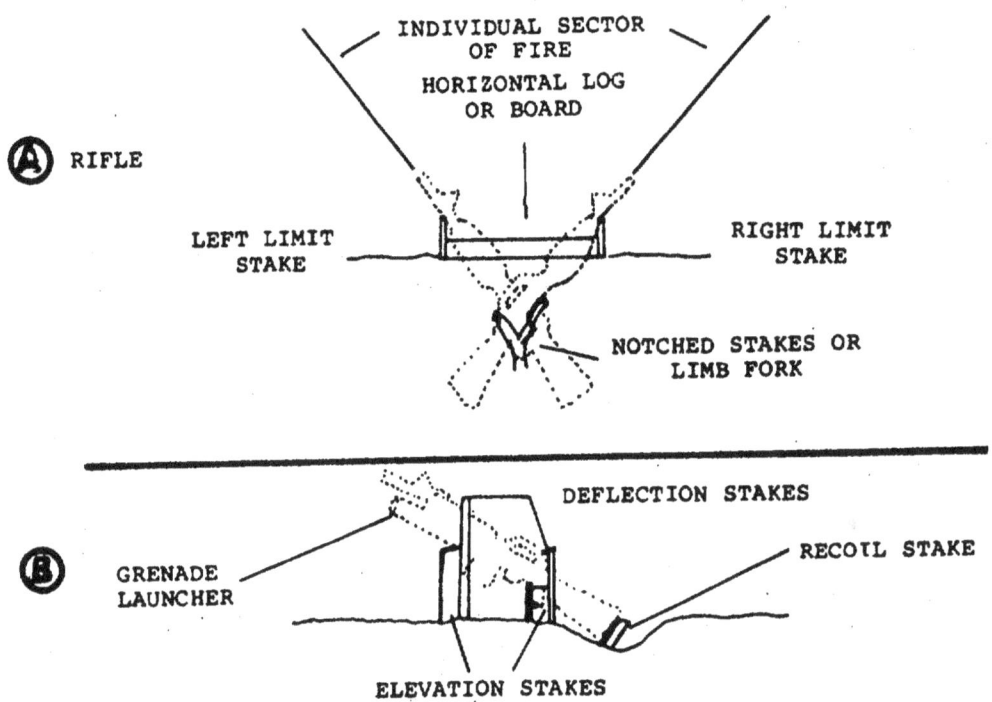

Figure 2-18. Field Expedients for Delivery of Preplanned Fires.

2605. Rate of Fire

The Marine is trained to fire approximately 10 to 12 aimed shots per minute (average rate). Difficulties encountered in battle usually make a slower rate advisable. *The fastest rate at which any rifleman or automatic rifleman should fire is determined by his ability to select targets, align the sights, and squeeze off accurate shots.*

The automatic rifle is particularly valuable against targets such as machine guns and automatic weapons. The rapid rate of fire for the automatic rifle is 100 rounds per minute. The sustained rate of fire is 85 rounds per minute. Determination of the rate of fire for the automatic rifle is governed by the nature of the target. When beginning a fire fight, the first few rounds of automatic rifle fire should be delivered at the rapid rate in order to gain fire superiority and to fix the enemy. Thereafter, the rate should be slowed to the sustained rate, which is normally sufficient to maintain fire superiority.

2606. Fire Control and Fire Discipline

In order for a unit's fire to be effective, the unit leader must exercise fire control. Fire control relates to the leader's ability to have his men open or cease fire at the instant he desires, to adjust fire onto a target, to shift all or part of the fire from one target to another, and to regulate the rate of fire. The leader must teach his men fire discipline so that he may exercise fire control. Fire discipline is achieved when the unit has been taught and pays strict attention to instructions regarding the use of the rifle, automatic rifle, and grenade launcher, and can collectively execute fire commands with precision.

The unit leader must supervise and control the fire of his men so that it is directed and maintained at suitable targets. Upon receipt of orders, commands, or signals from the platoon commander, a squad leader promptly orders his squad to perform the fire mission directed. He is normally located at the rear of his squad during a fire fight. He usually gives his orders to the squad through the fire team leaders, but he does whatever is necessary to control the fire of his squad effectively. Squad and fire team leaders exercise fire control by means of voice commands and signals.

Chapter 3

Combat Formations and Signals

Section I. Combat Formations

3101. General

Fire team and squad combat formations are groupings of individuals and units for efficient tactical employment. The factors influencing the leader's decision as to the selection of a particular formation are the mission, terrain, situation, weather, speed, and degree of flexibility. Combat formations and signals enable the leader to control the fire and maneuver of his unit when moving to and assaulting an enemy position.

3102. Basic Combat Formations

a. Fire Team. Normally each fire team leader will determine the formation for his own unit. Thus, a squad may contain a variety of fire team formations at any one time and these formations may change frequently. The relative position of the fire teams within the squad formation should be such that one will not mask the fire of the others. It is not important that exact distances and intervals be maintained between fire teams and individuals as long as control is not lost. Sight or voice contact will be maintained within the fire team and between fire team leaders and squad leaders. All movement incident to changes of formation is usually by the shortest practical route. The characteristics of fire team formations are similar to those of corresponding squad formations. The characteristics of the fire team formations are as follows: (See figs. 3-1 through 3-4.)

(1) **Column**

(a) Permits rapid, controlled movement.

(b) Favors fire and maneuver to the flanks.

(c) Vulnerable to fire from the front and provides the least amount of fire to the front.

3-2

Figure 3-1. Fire Team Column.

(2) Wedge

    (a) Permits good control.

    (b) Provides all-round security.

    (c) Formation is flexible.

    (d) Fire is adequate in all directions.

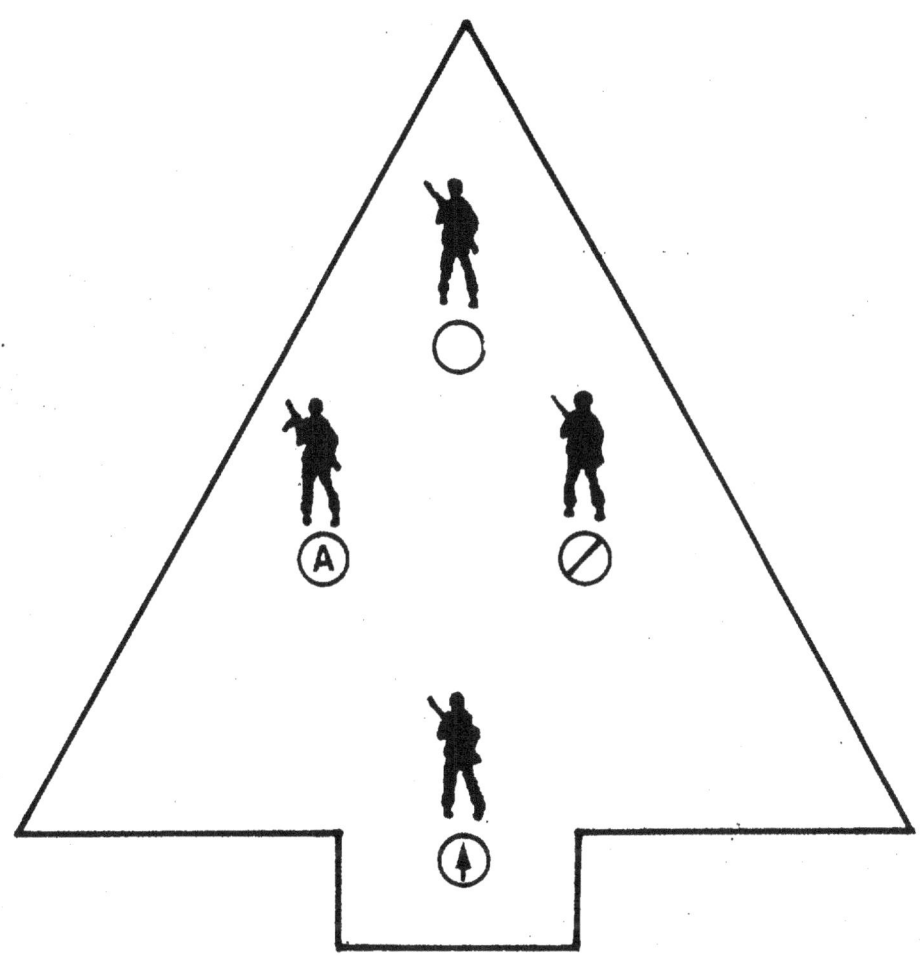

NOTE: THE POSITION OF THE FIRE TEAM LEADER AND ASSISTANT AUTOMATIC RIFLEMAN ARE INTERCHANGEABLE.

Figure 3-2. Fire Team Wedge.

(3) **Skirmishes Right (Left)**

(a) Maximum firepower to the front.

(b) Used when the location and strength of the enemy are known, during the assault, mopping up, and crossing short open areas.

Figure 3-3. Fire Team Skirmishers.

(4) Echelon Right (Left)

(a) Provides heavy firepower to front and echeloned flank.

(b) Used to protect an open or exposed flank.

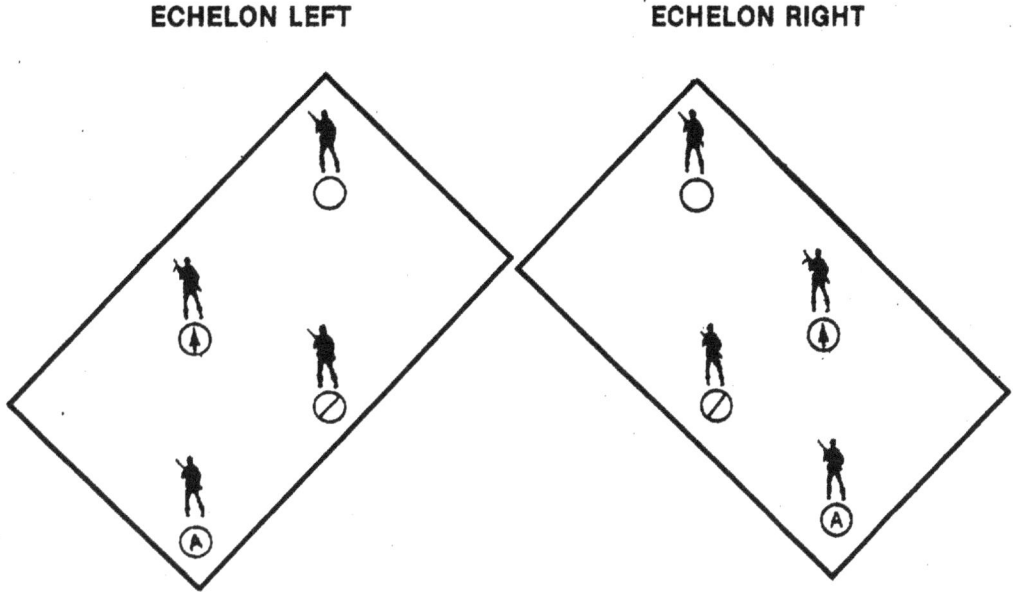

Figure 3-4. Fire Team Echelon.

b. Squad. The squad leader prescribes the formation for his squad. The platoon leader and squad leader may prescribe the initial formation for their respective subordinate units when the situation dictates or the commander so desires. Subsequent changes may be made by the subordinate unit leaders. The characteristics of squad formations are similar to those of the fire team. The fire team is the maneuver element in squad formations. (See figs. 3-5 through 3-11.)

(1) **Squad Column.** Fire teams are arranged in succession one behind the other.

(a) Easy to control and maneuver.

(b) Excellent for speed of movement or when strict control is desired.

(c) Especially suitable for narrow covered routes of advance, maneuvering through gaps between areas receiving hostile artillery fire, moving through areas of limited observation, and moving under conditions of reduced visibility.

3-5

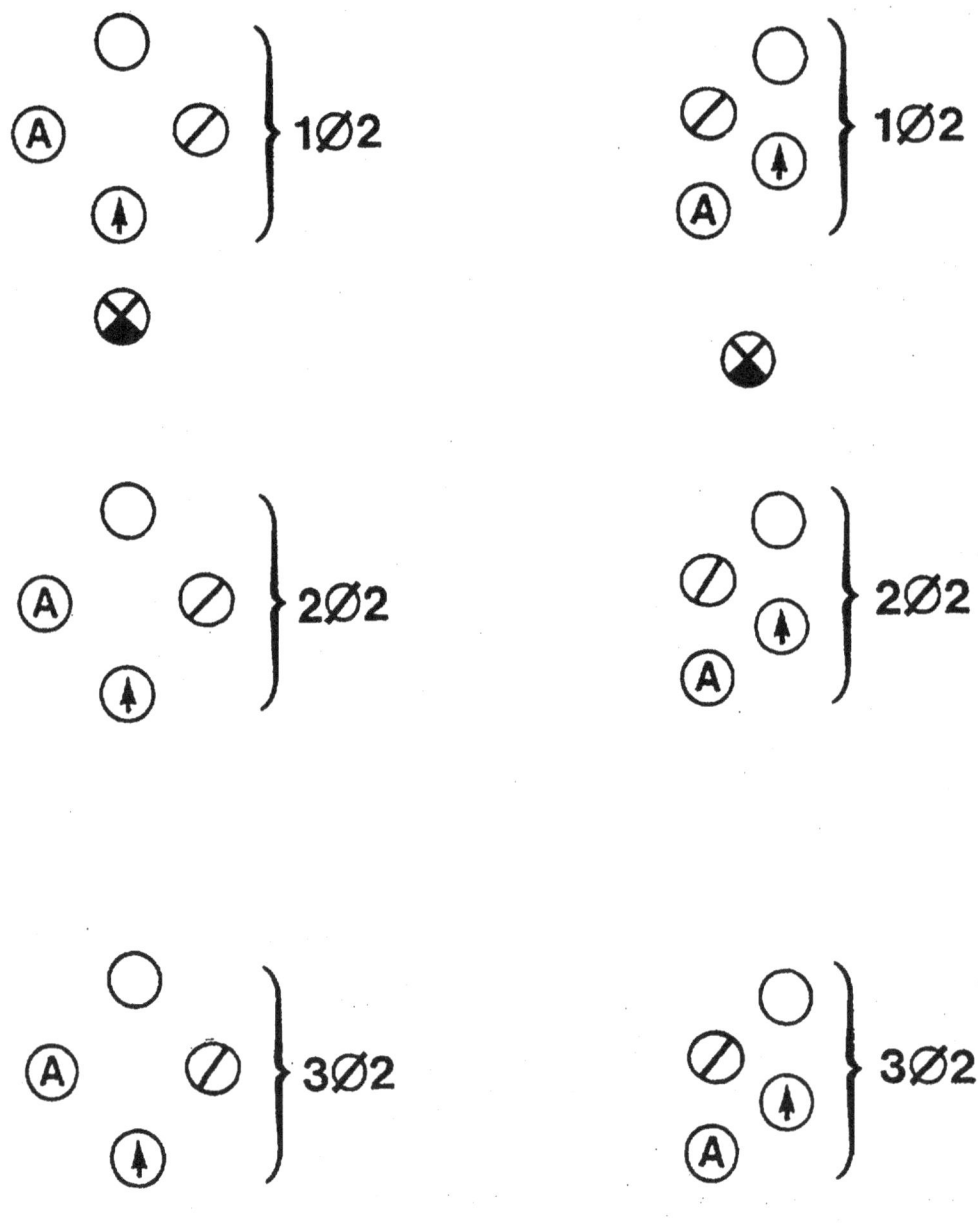

FIRE TEAMS IN WEDGE FIRE TEAMS IN COLUMN

Figure 3-5. Squad Column.

(d) Vulnerable to fire from the front.

(e) Used for night operations.

(2) Squad Wedge. See discussion under fire team formations for the wedge in paragraph 3102a(2).

(3) Squad Vee

(a) Facilitates movement into squad line.

(b) Provides excellent firepower to front and flanks.

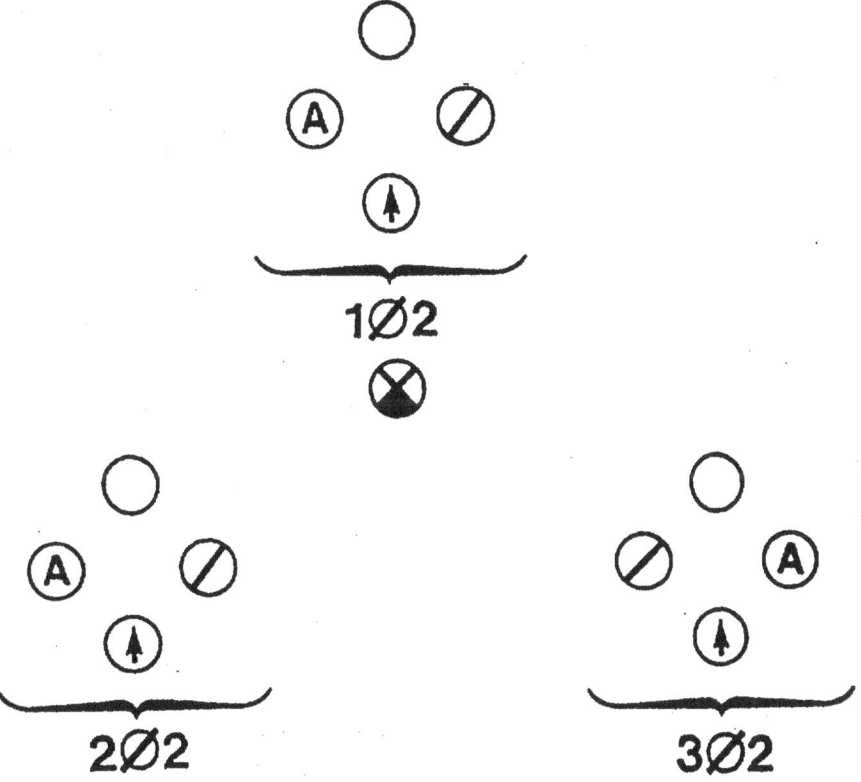

(FIRE TEAMS IN WEDGE. TEAM LEADERS POSITIONED FOR
EASE IN COMMUNICATING WITH SQUAD LEADER.)

Figure 3-6. Squad Wedge.

(c) Provides all-round security.

(d) Used when the enemy is to the front and his strength and location are known. May be used when crossing large open areas.

(4) Squad Line. See discussion under fire team formations for skirmishers right (left) in paragraph 3102a(3).

(5) Squad Echelon. See discussion under fire team formations for echelons right (left) in paragraph 3102a(4).

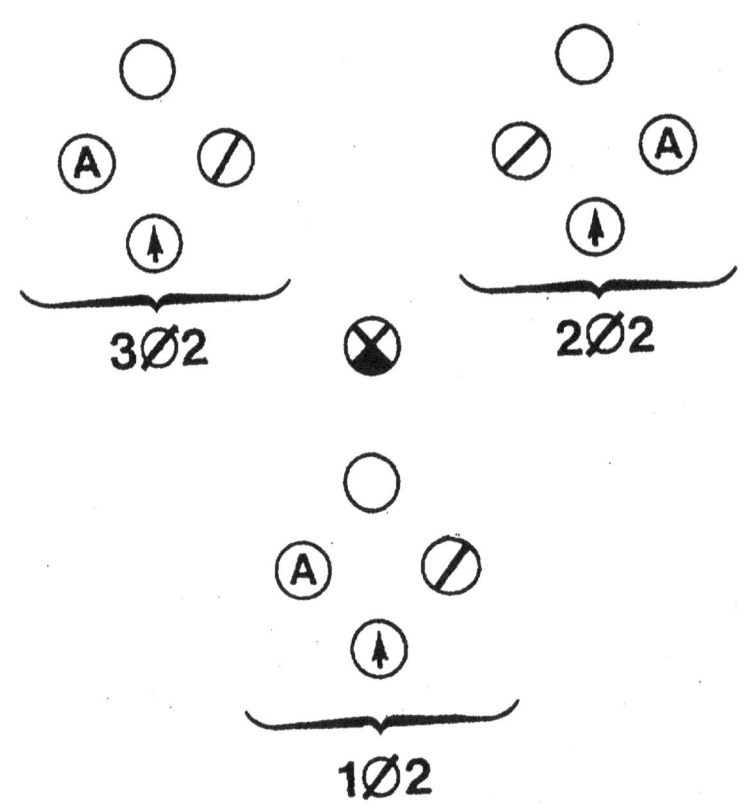

(FIRE TEAMS IN WEDGE. TEAM LEADERS POSITIONED FOR EASE IN COMMUNICATING WITH SQUAD LEADER.)

Figure 3-7. Squad Vee.

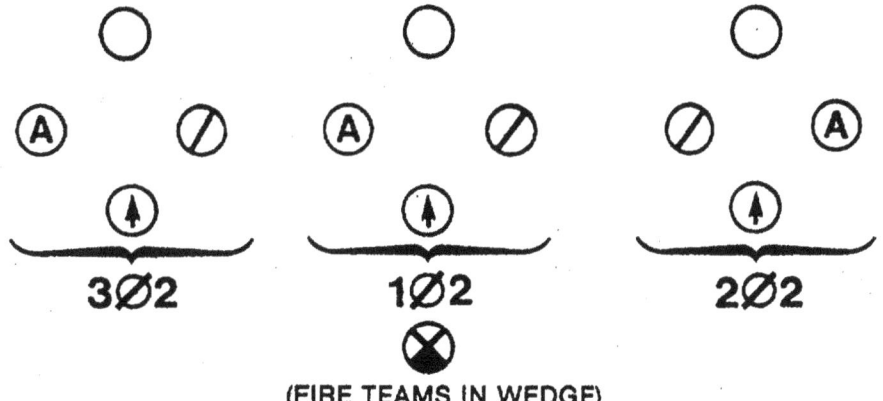

(FIRE TEAMS IN WEDGE)

Figure 3-8. Squad Line.

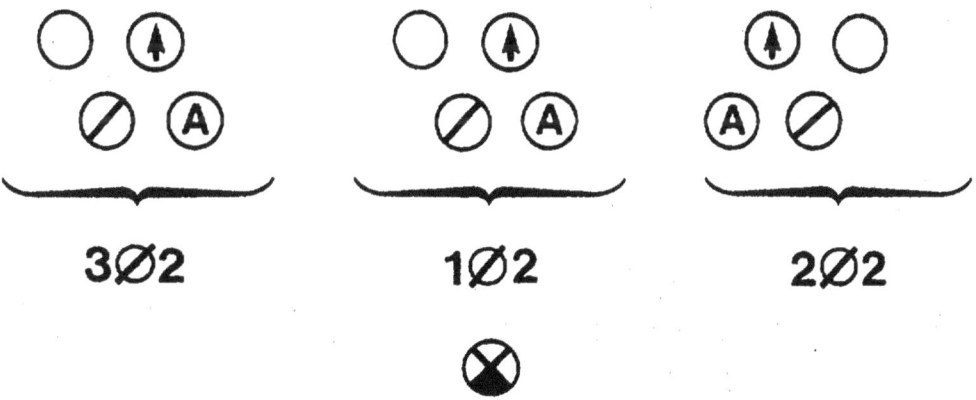

(FIRST AND THIRD FIRE TEAMS IN SKIRMISHERS RIGHT: SECOND FIRE TEAM IN SKIRMISHERS LEFT)

Figure 3-9. Squad Line.

3-10

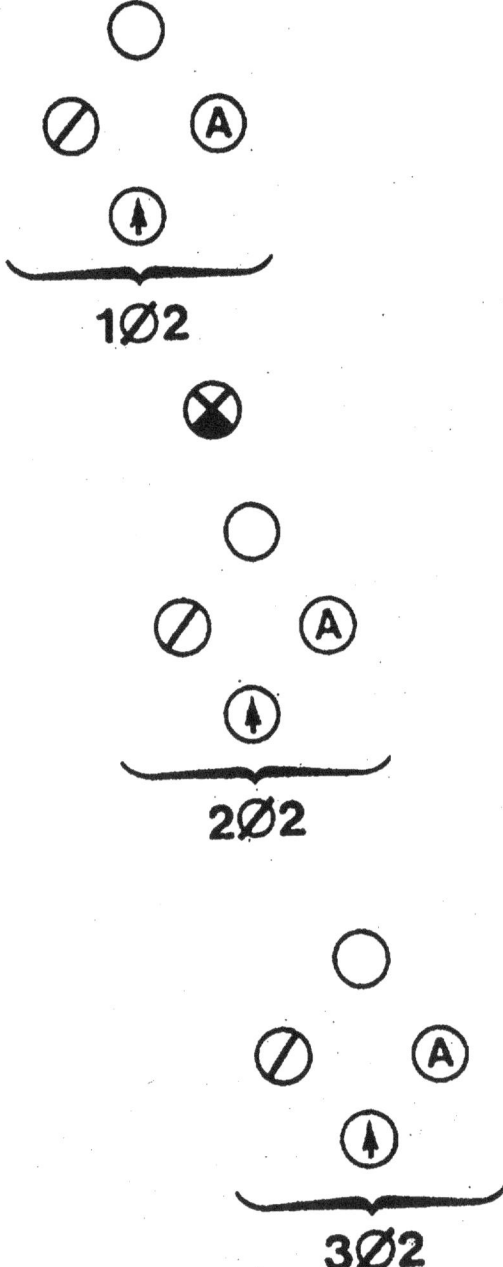

Figure 3-10. Squad Echelon Right.

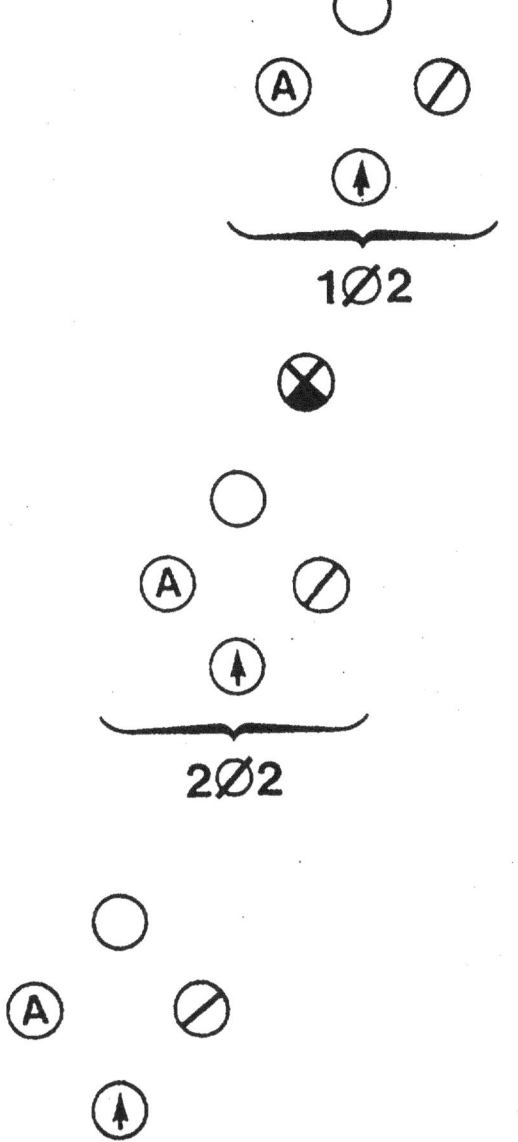

(FIRE TEAMS IN WEDGE)

Figure 3-11. Squad Echelon Left.

3103. Changing Formations (Battle Drill)

The squad leader may change formations to reduce casualties from hostile fire, present a less vulnerable target, or get over difficult or exposed terrain. Formation changes in varying or rough terrain are frequent in order to get the squad over manmade obstacles and natural obstacles such as rivers, swamps, jungles, woods, and sharp ridges.

Directions of movement for members of the fire team when the leader signals for changes of formation are shown in figure 3-12 a through u. Figure 3-12 is provided only as a guide to assist fire teams in developing the most rapid means of moving from one formation to another. When the team is to execute a movement, the fire team leader signals with his arms and hands, indicating the movement and direction.

The squad leader signals the squad formation to the fire team leaders. Remember, the fire team may be in any fire team formation within the squad formation.

BATTLE DRILL FIRE TEAM FORMATIONS

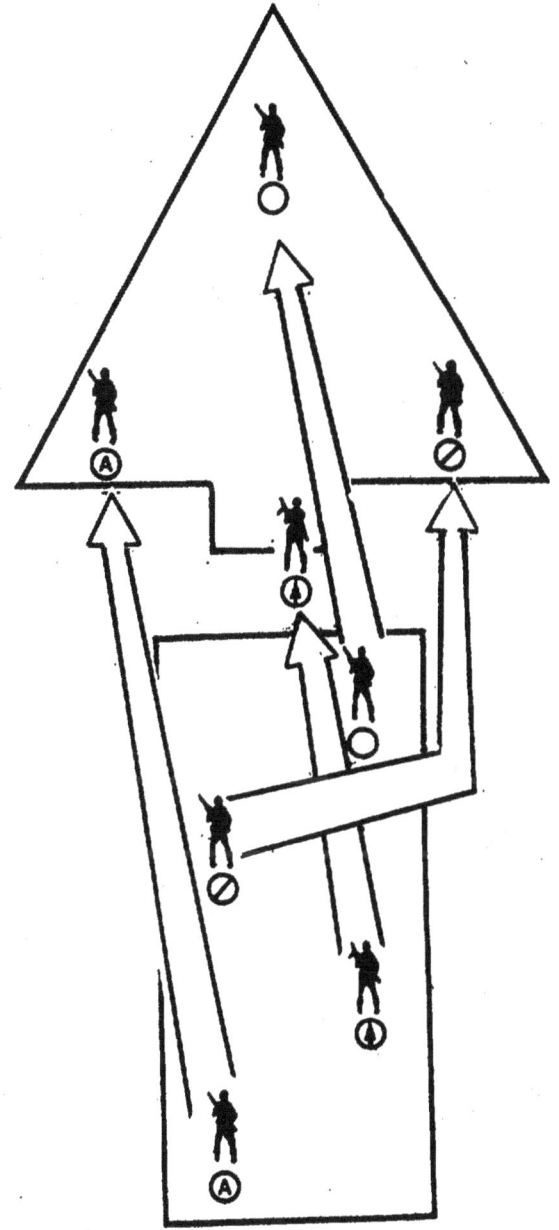

COLUMN TO WEDGE

Figure 3-12 a. Changing Formations.

3-14

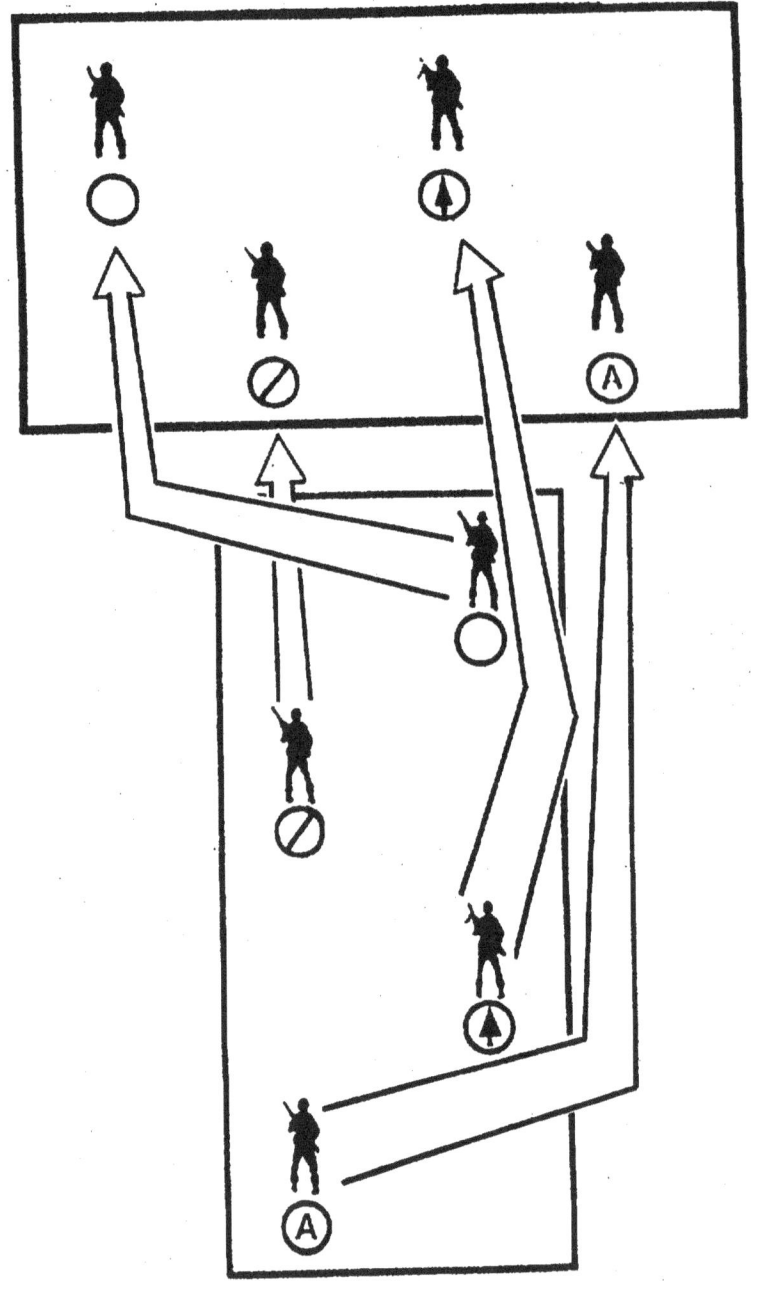

COLUMN TO SKIRMISHERS RIGHT

Figure 3-12 b. Changing Formations. (Continued)

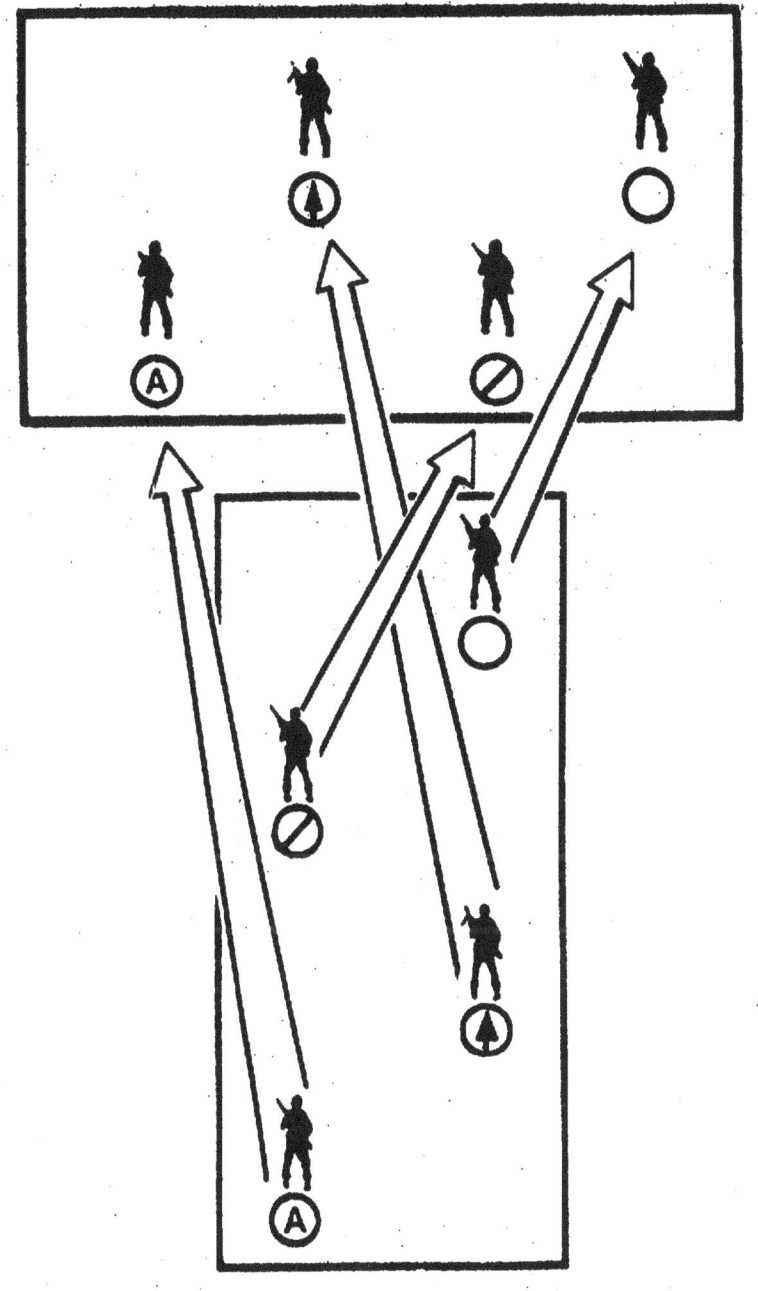

COLUMN TO SKIRMISHERS LEFT

Figure 3-12 c. Changing Formations. (Continued)

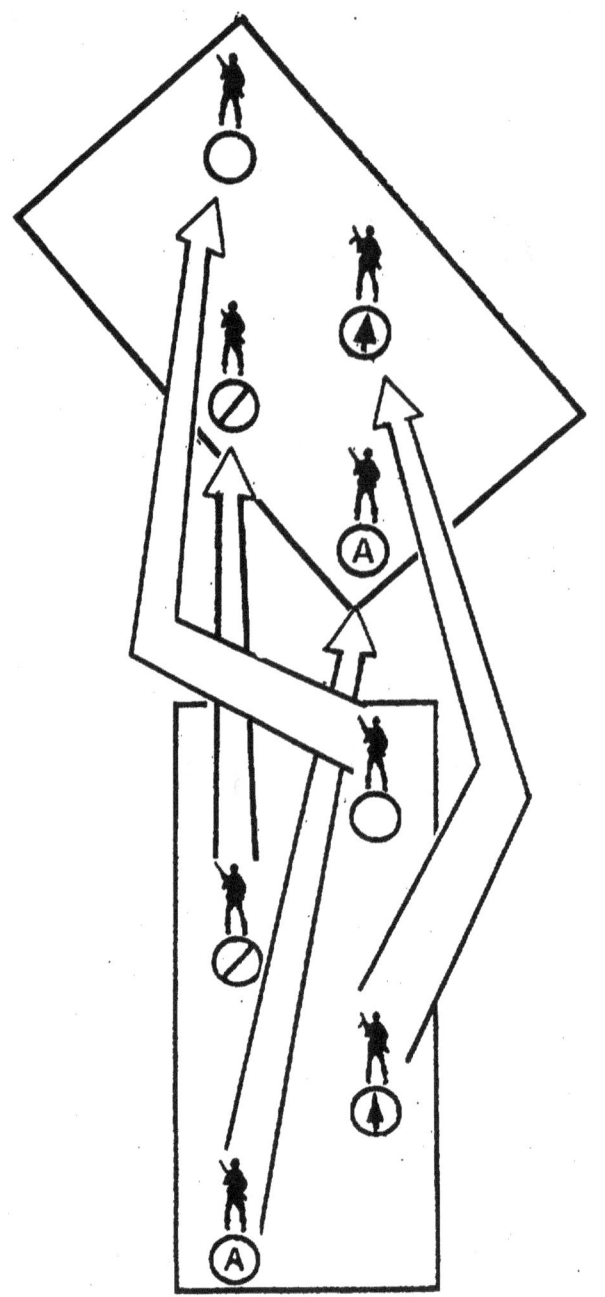

COLUMN TO ECHELON RIGHT

Figure 3-12 d. Changing Formations. (Continued)

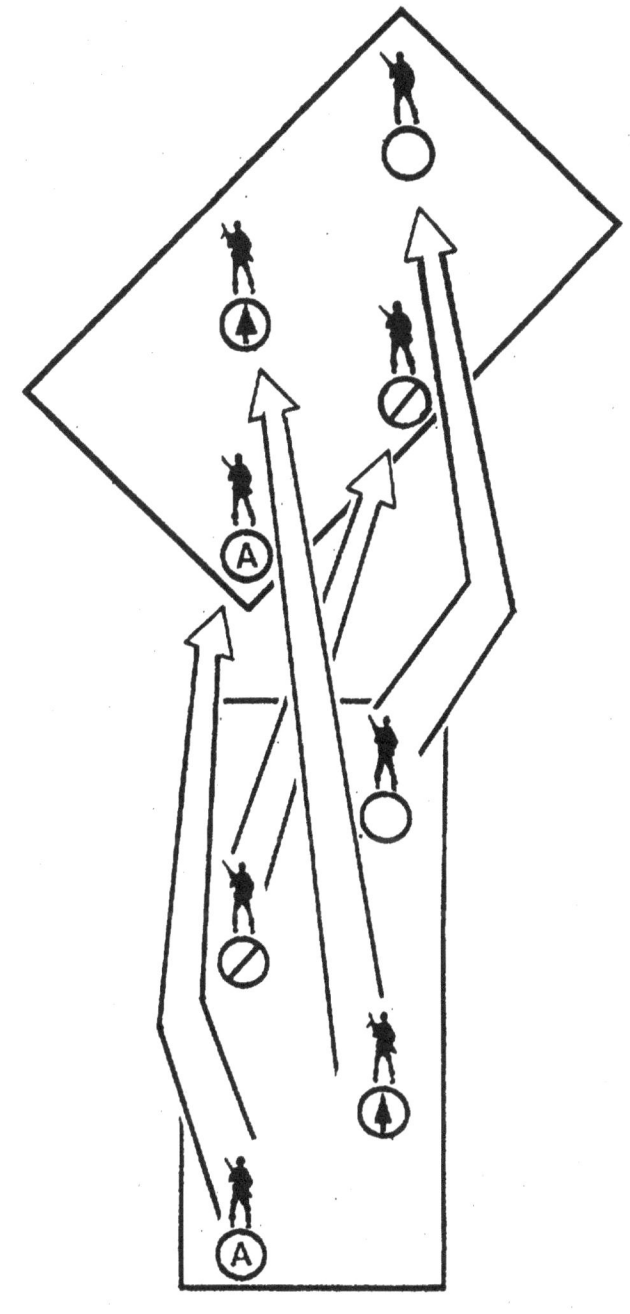

COLUMN TO ECHELON LEFT

Figure 3-12 e. Changing Formations. (Continued)

3-18

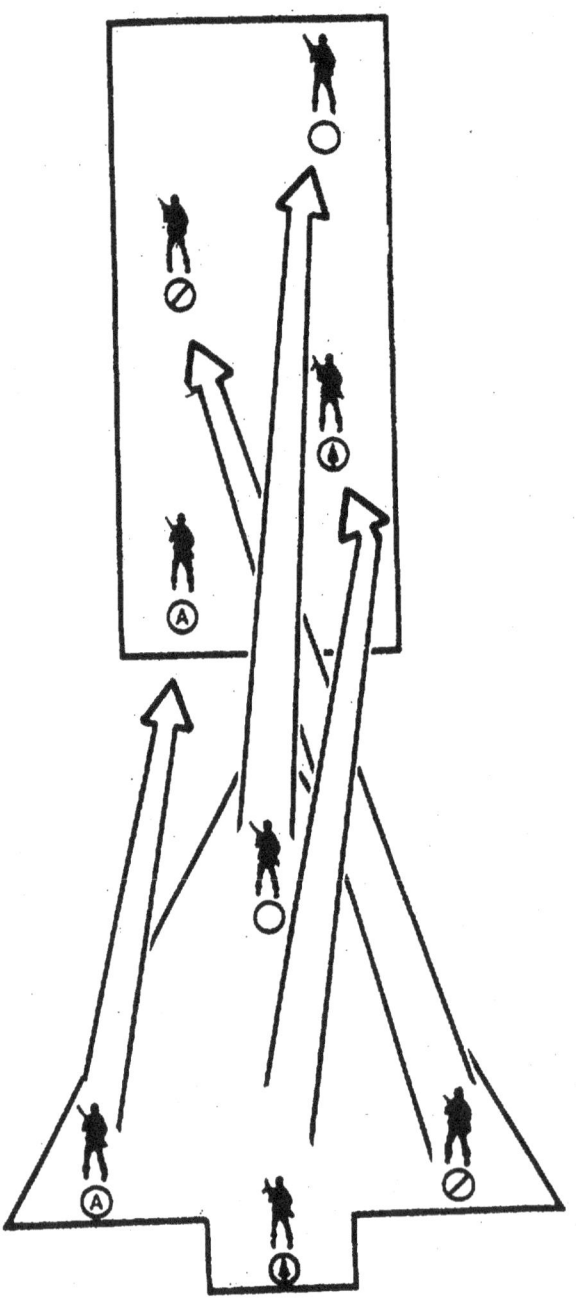

WEDGE TO COLUMN

Figure 3-12 f. Changing Formations. (Continued)

SKIRMISHERS RIGHT

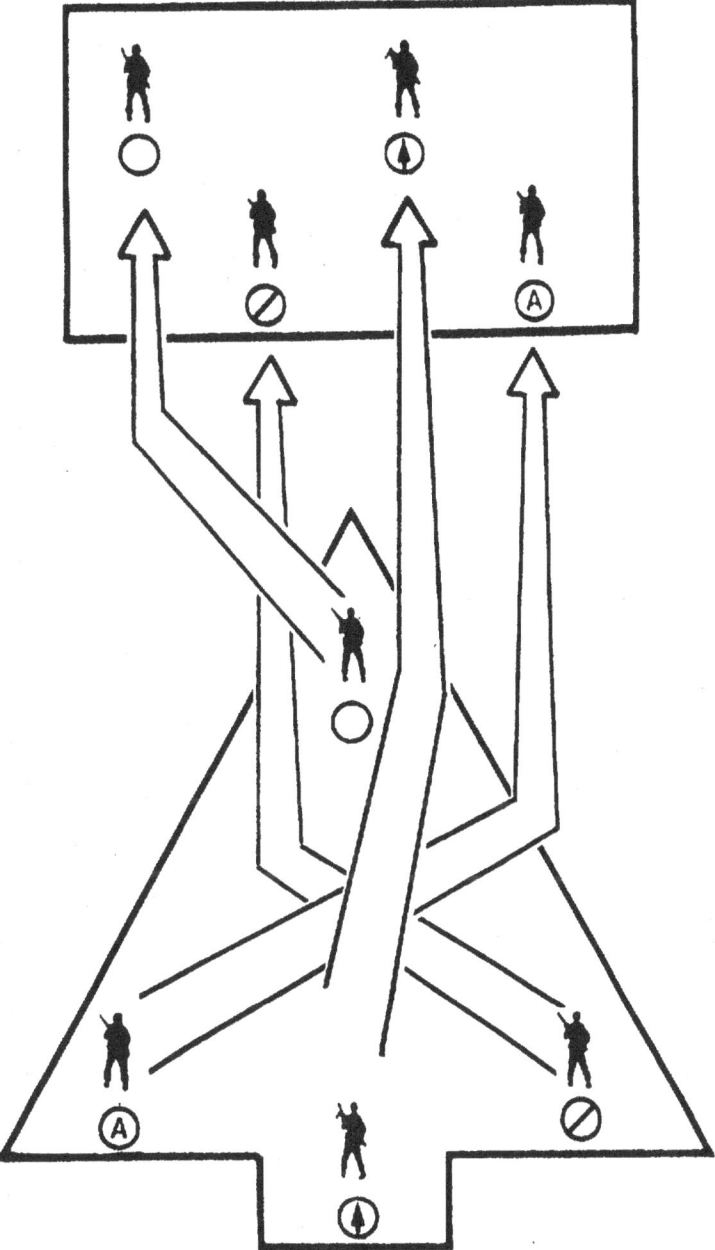

WEDGE TO SKIRMISHERS RIGHT

Figure 3-12 g. Changing Formations. (Continued)

3-20

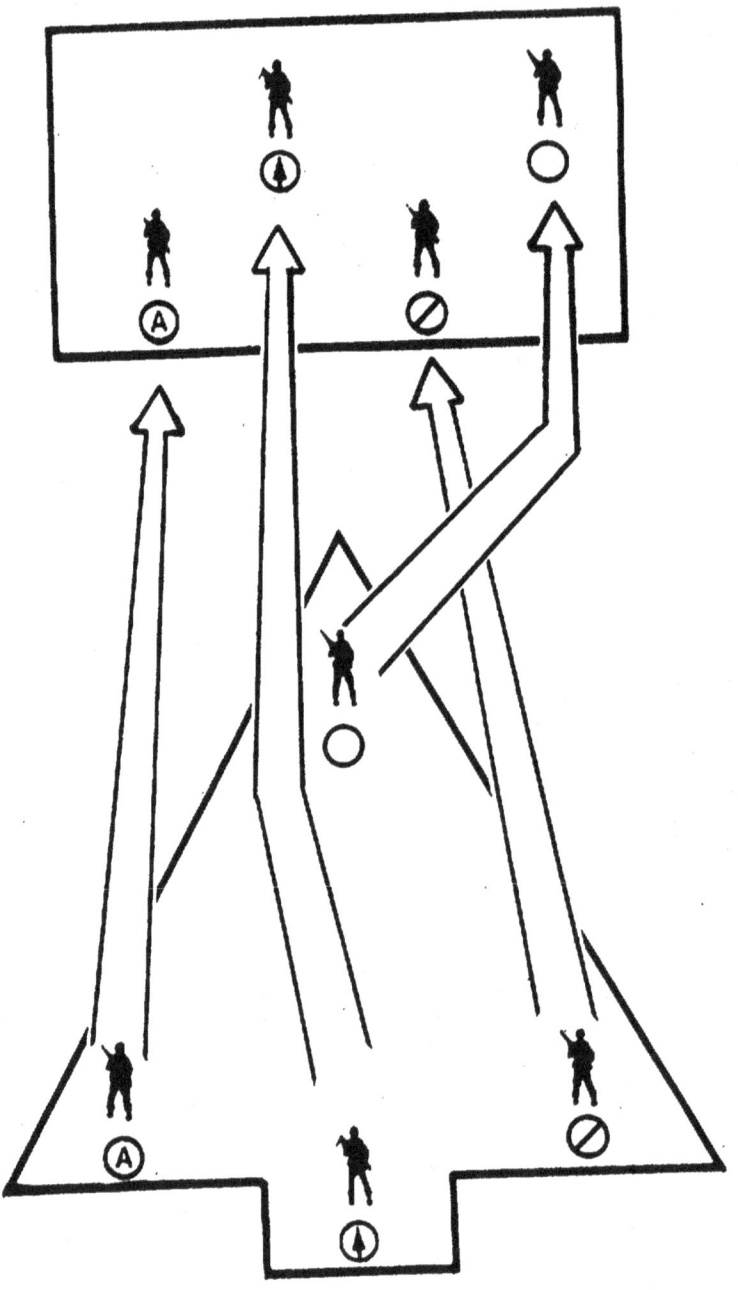

WEDGE TO SKIRMISHERS LEFT

Figure 3-12 h. Changing Formations. (Continued)

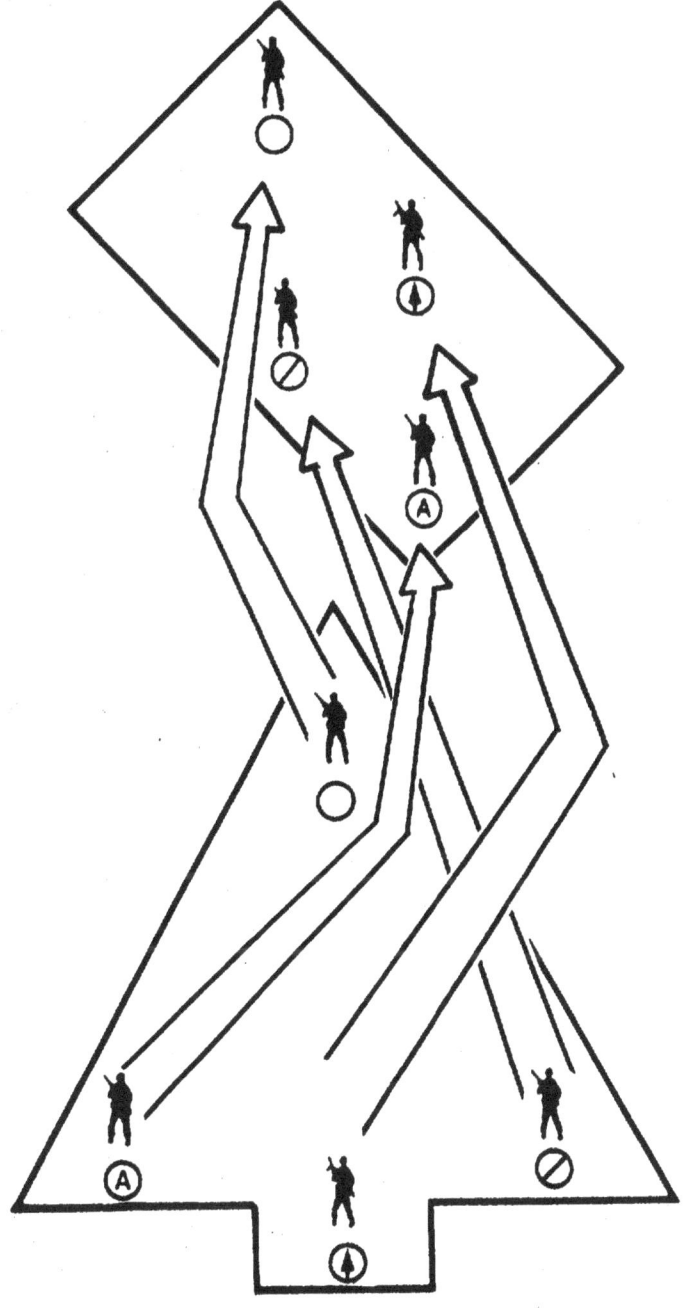

WEDGE TO ECHELON RIGHT

Figure 3-12 l. Changing Formations. (Continued)

3-22

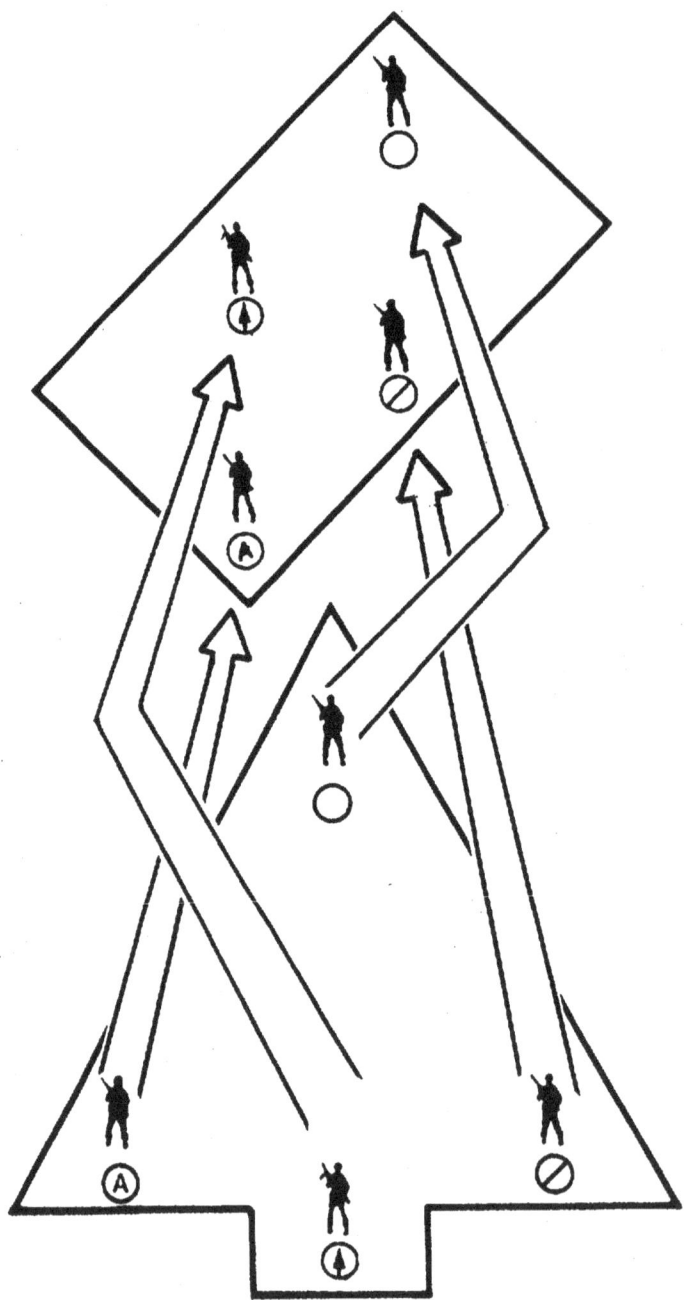

WEDGE TO ECHELON LEFT

Figure 3-12 j. Changing Formations. (Continued)

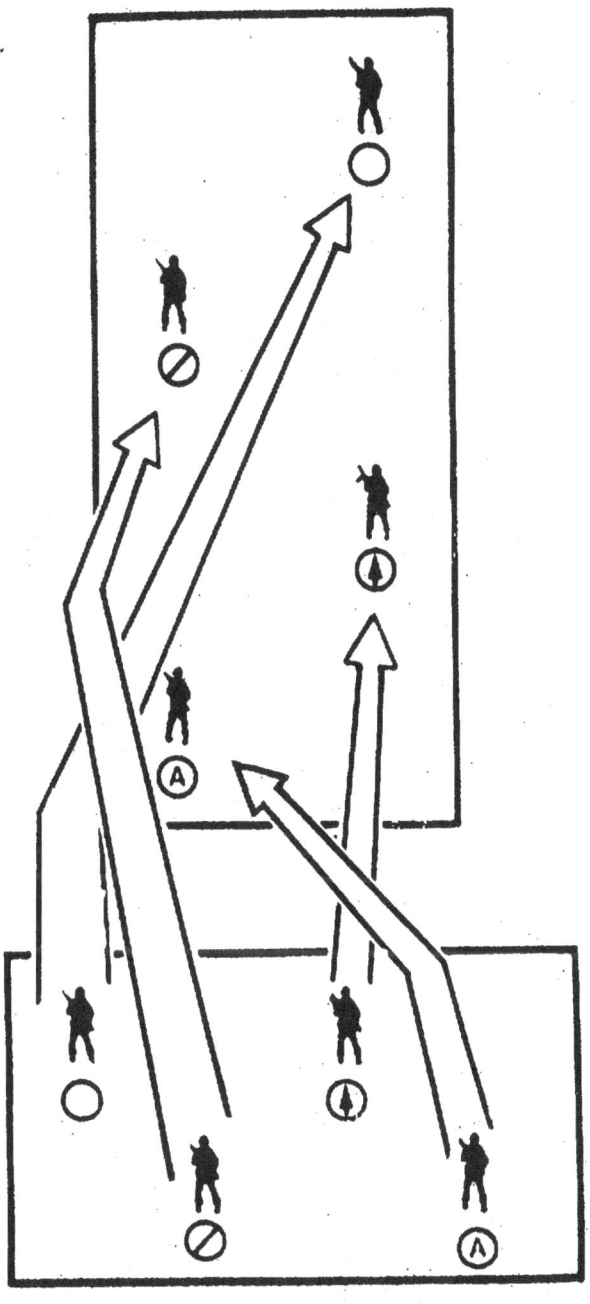

SKIRMISHERS RIGHT TO COLUMN

Figure 3-12 k. Changing Formations. (Continued)

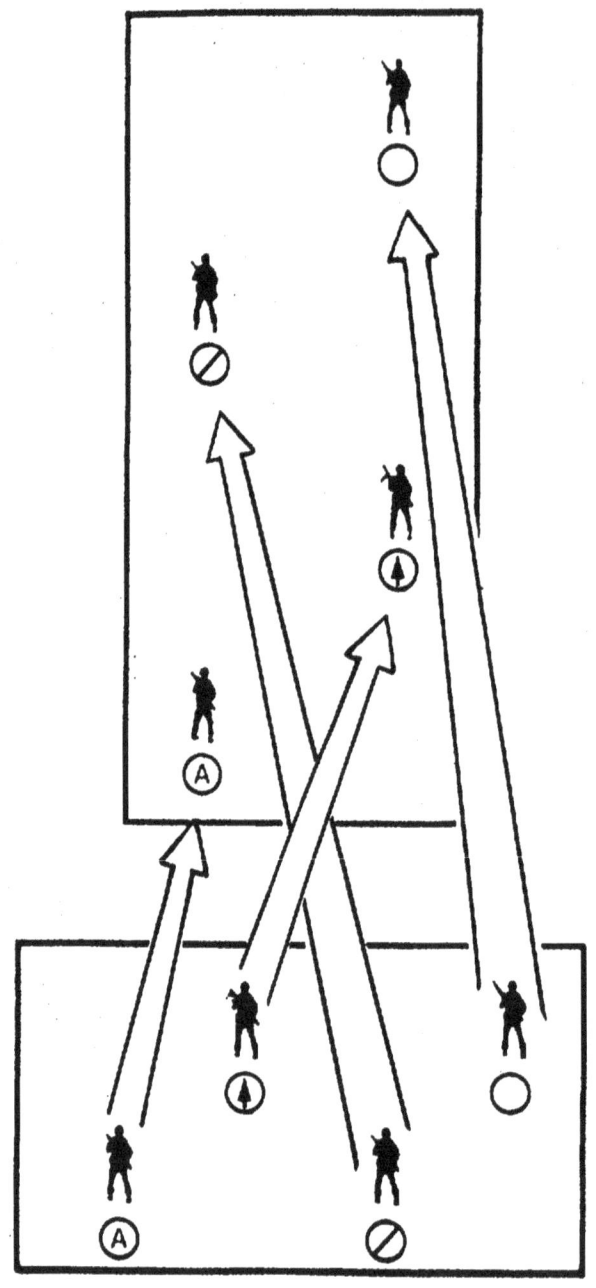

SKIRMISHERS LEFT TO COLUMN

Figure 3-12 I. Changing Formations. (Continued)

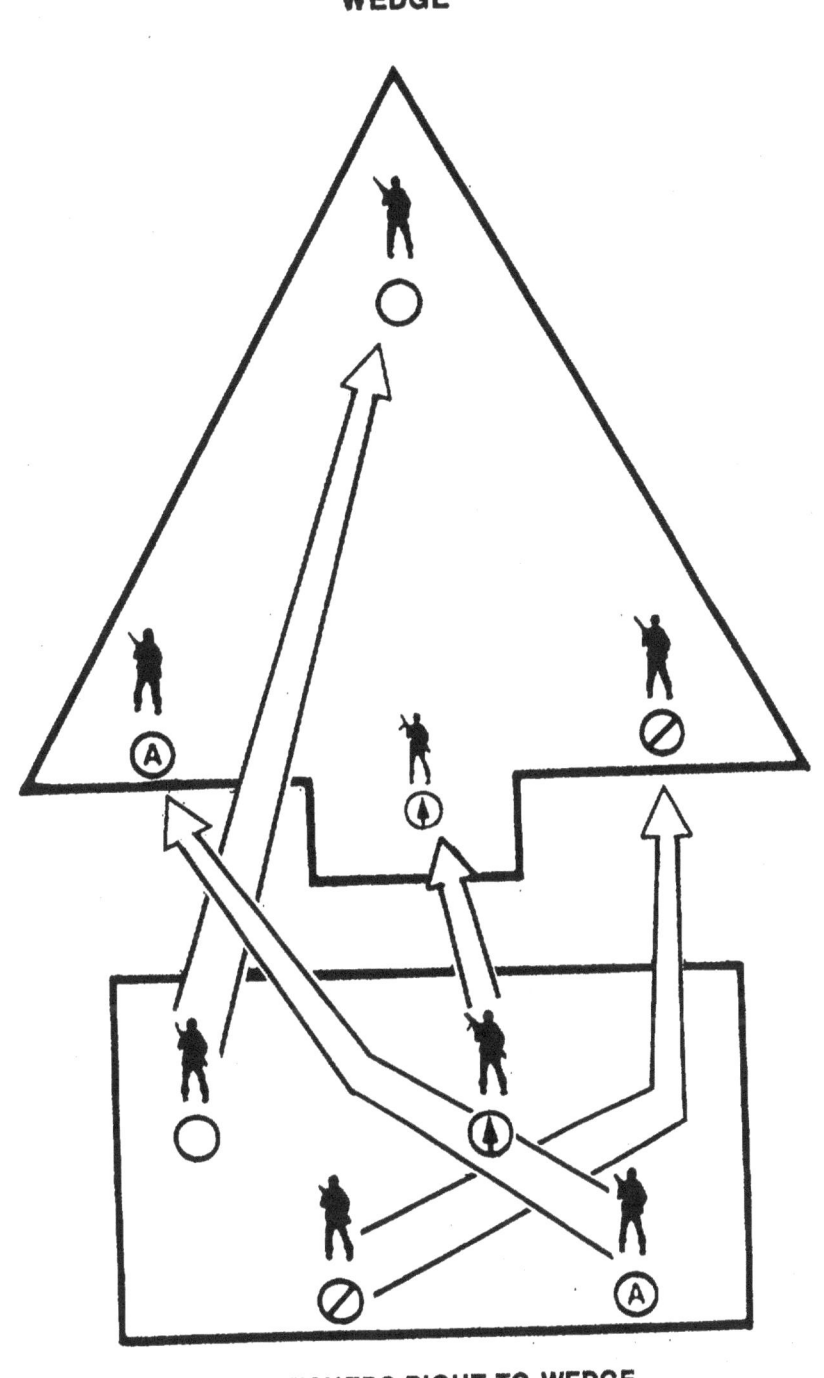

Figure 3-12 m. Changing Formations. (Continued)

3-26

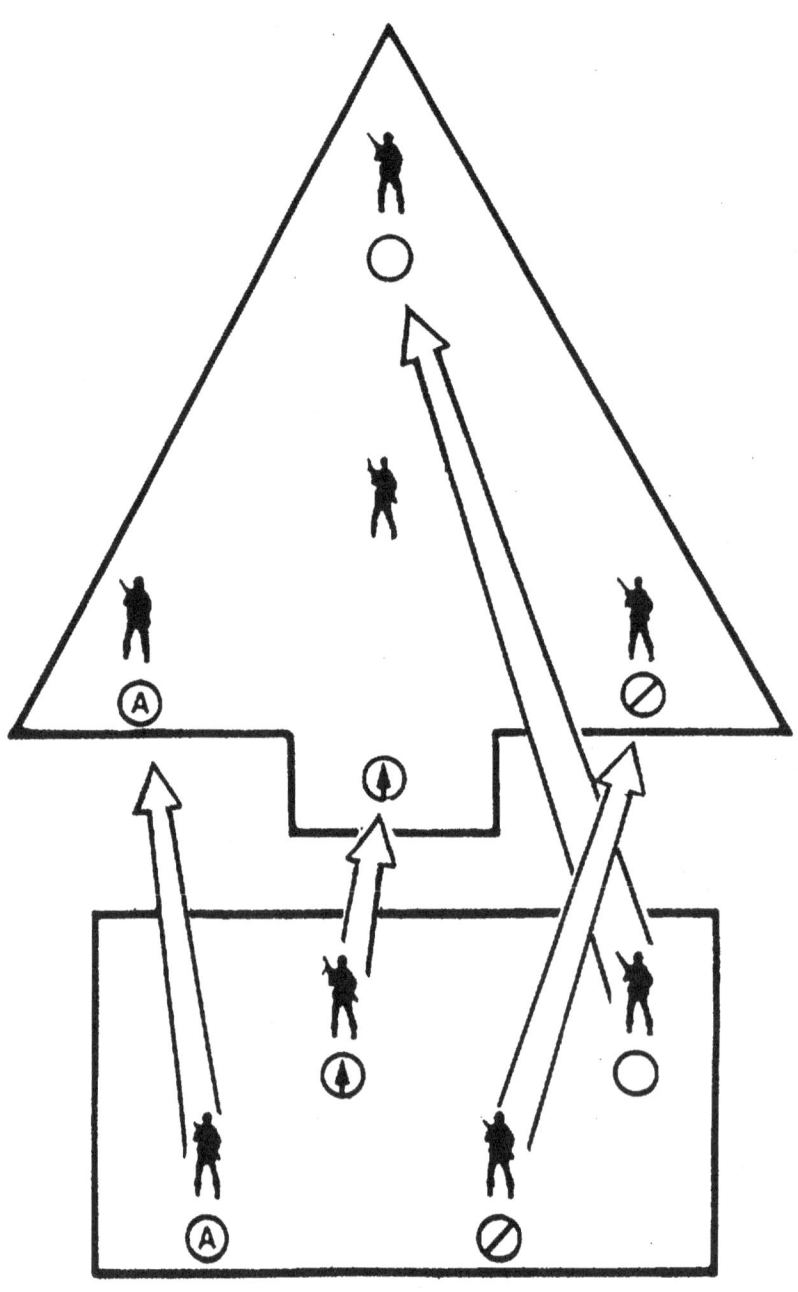

SKIRMISHERS LEFT TO WEDGE

Figure 3-12 n. Changing Formations. (Continued)

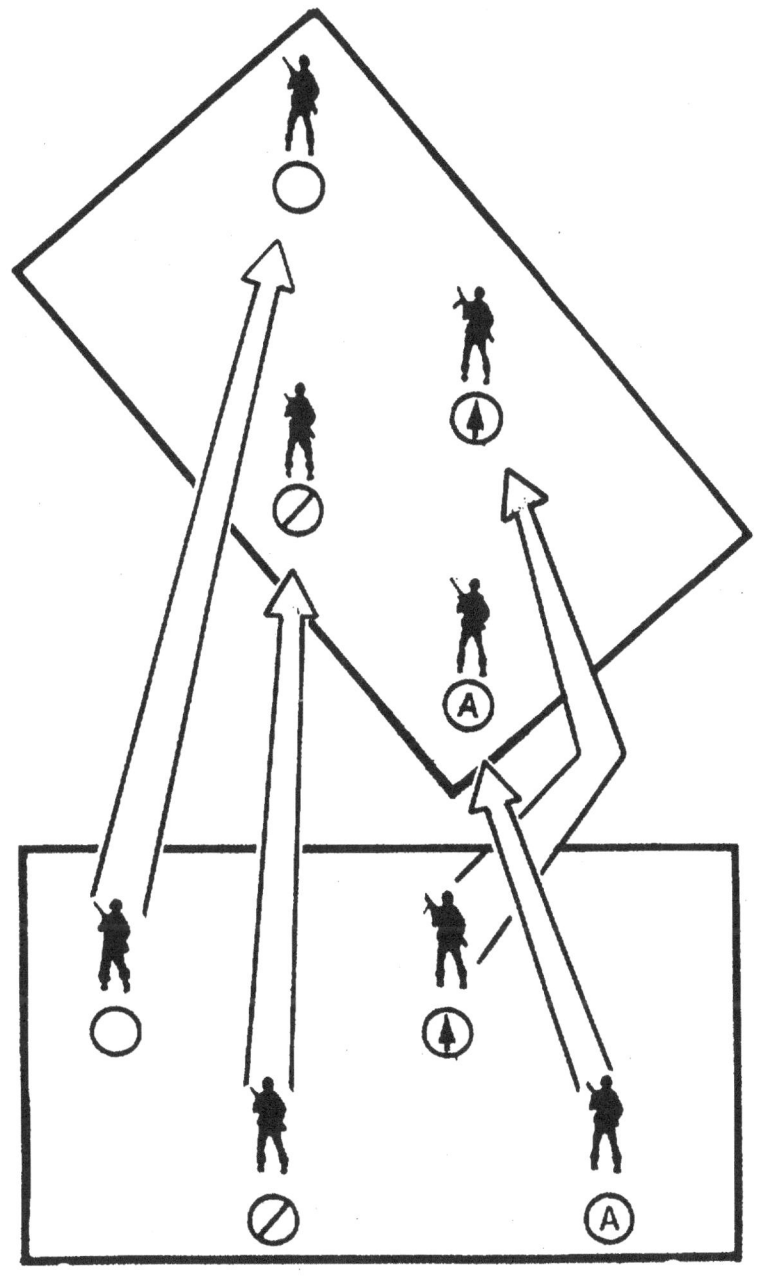

SKIRMISHERS RIGHT TO ECHELON RIGHT

Figure 3-12 o. Changing Formations. (Continued)

3-28

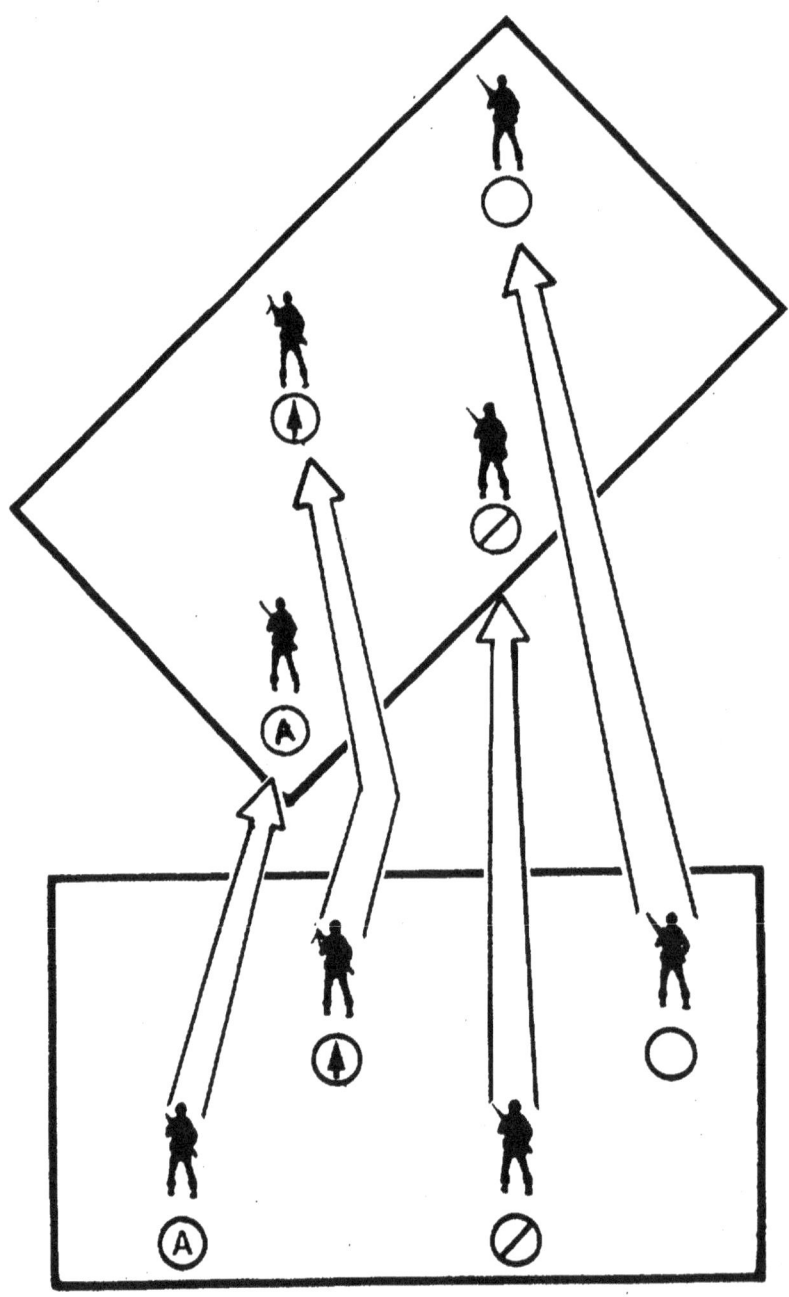

SKIRMISHERS LEFT TO ECHELON LEFT

Figure 3-12 p. Changing Formations. (Continued)

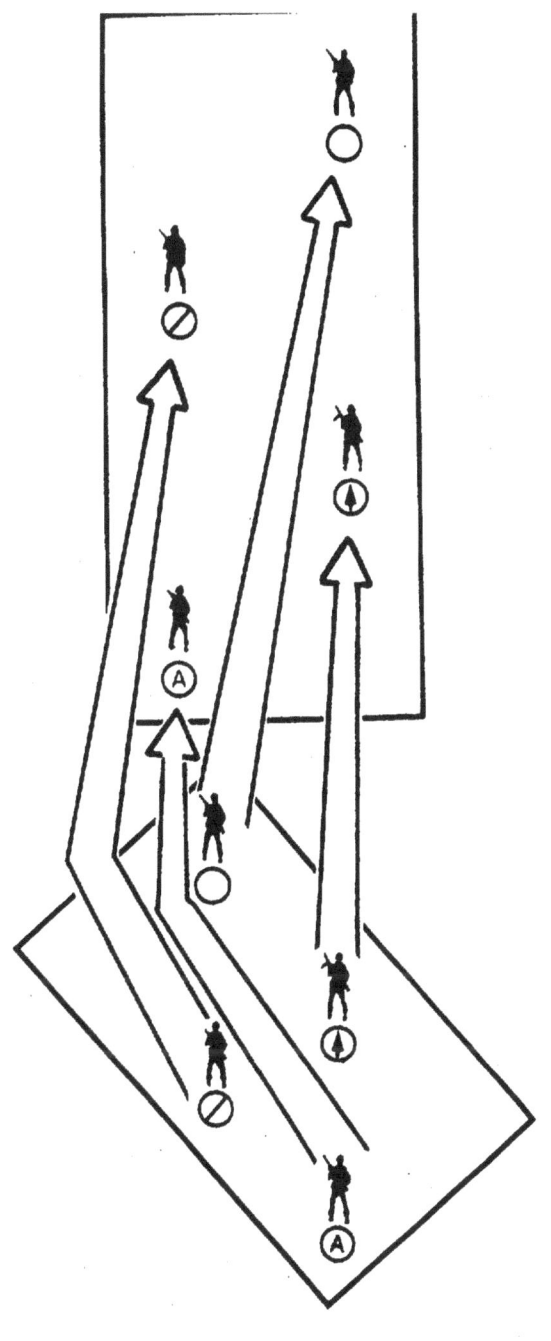

ECHELON RIGHT TO COLUMN

Figure 3-12 q. Changing Formations. (Continued)

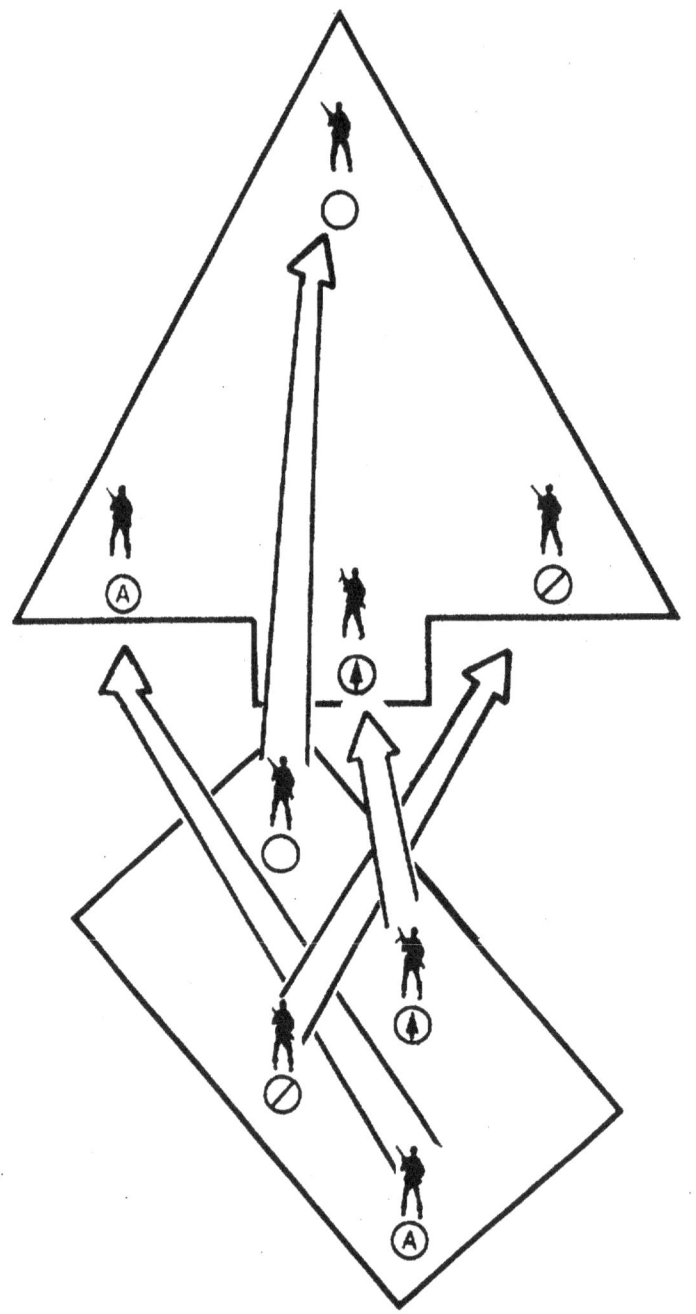

ECHELON RIGHT TO WEDGE

Figure 3-12 r. Changing Formations. (Continued)

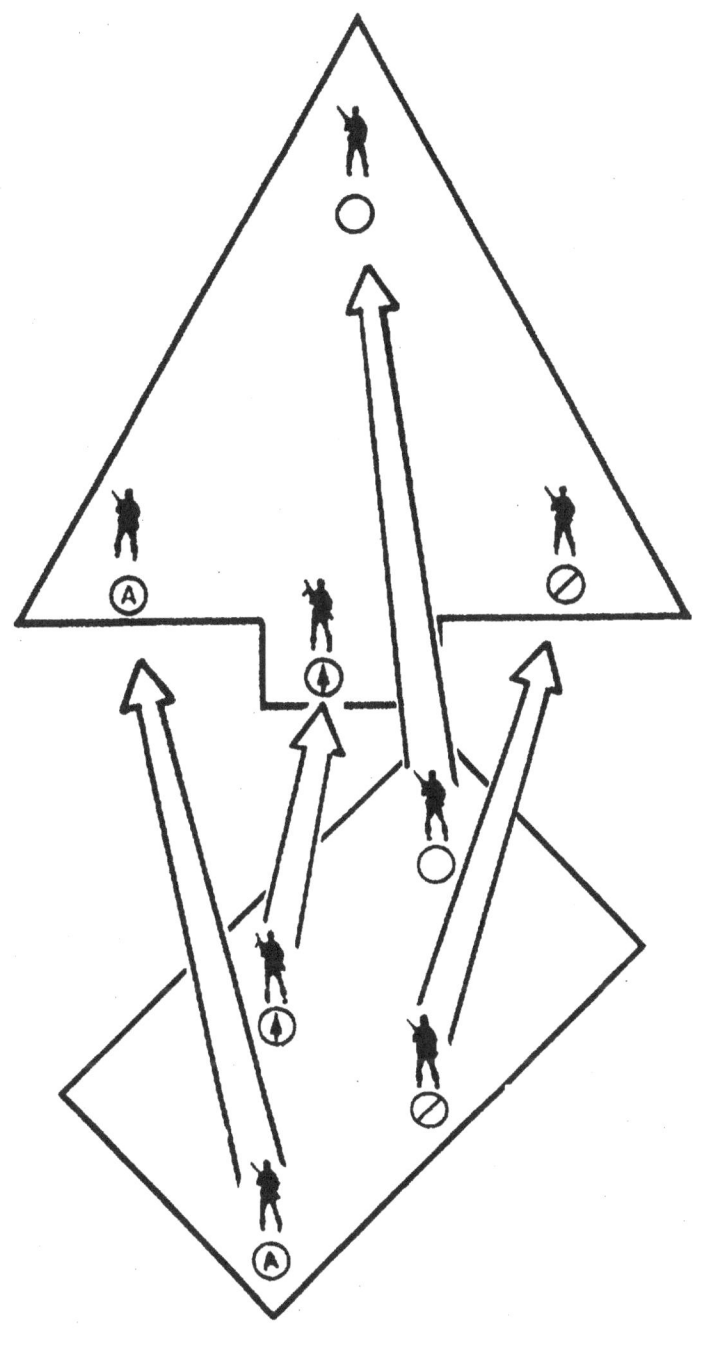

ECHELON LEFT TO WEDGE

Figure 3-12 s. Changing Formations. (Continued)

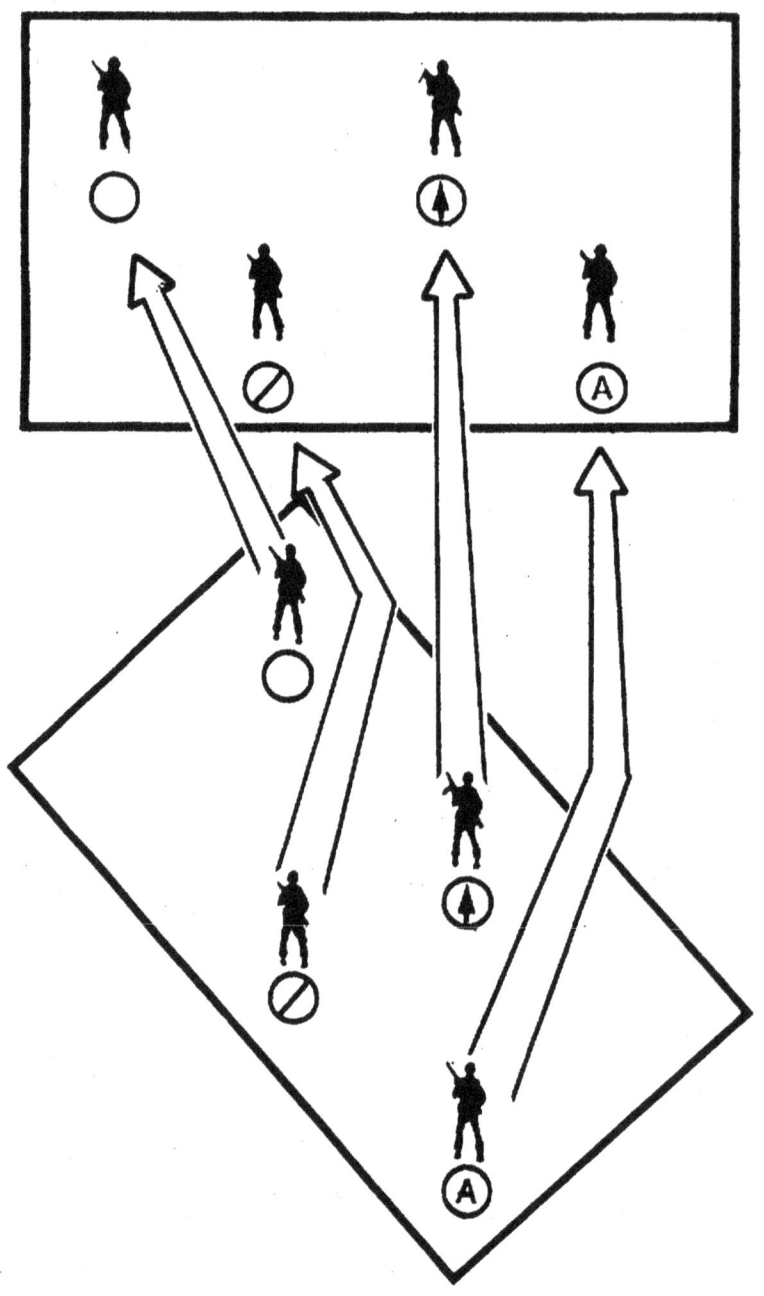

ECHELON RIGHT TO SKIRMISHERS RIGHT

Figure 3-12 t. Changing Formations. (Continued)

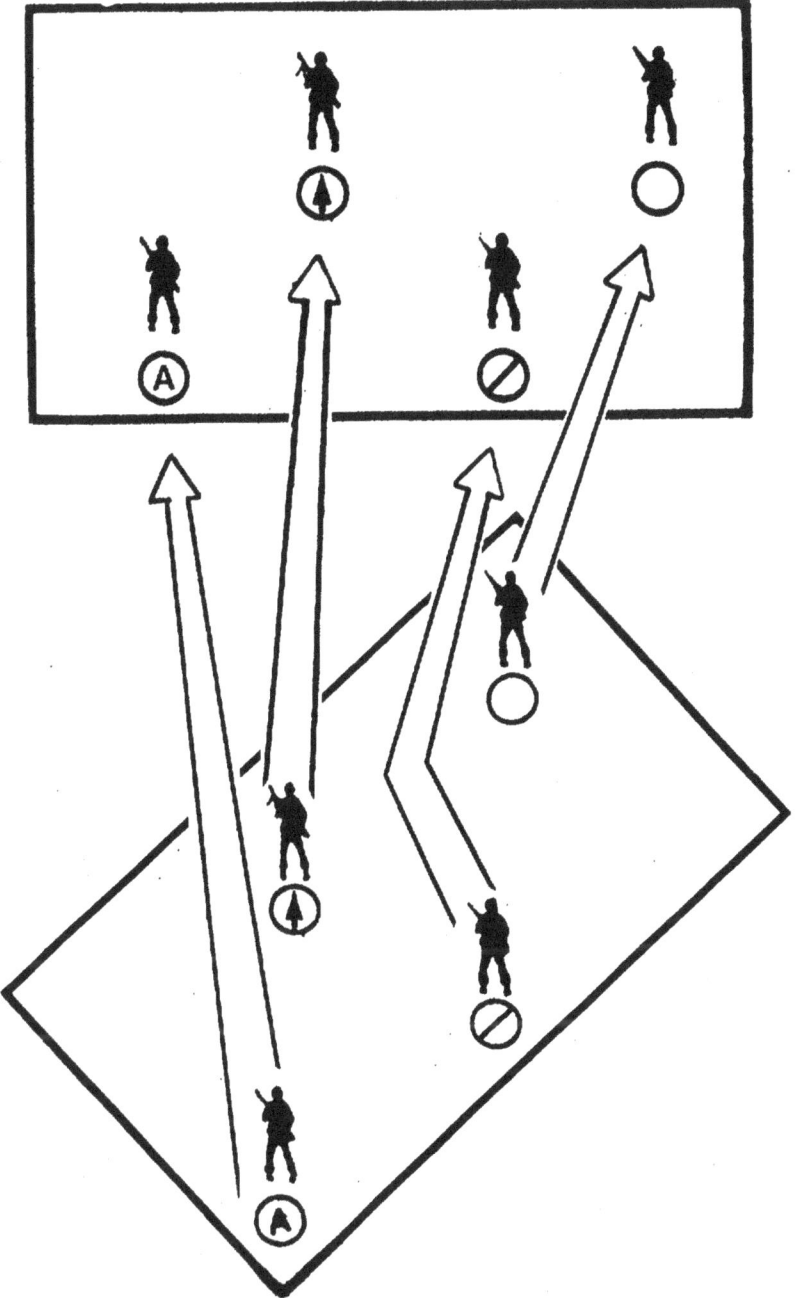

ECHELON LEFT TO SKIRMISHERS LEFT

Figure 3-12 u. Changing Formations. (Continued)

Section II. Signals

3201. General

Signals are used to transmit commands or information when voice communications are difficult, impossible, or when silence must be maintained. Subordinate leaders repeat signals to their units whenever necessary to ensure prompt and correct execution.

3202. Whistle

The whistle is an excellent signal device for the small unit leader. It provides a fast means of transmitting a message to a large group. However, unless the signal is prearranged and understood, it may be misinterpreted, and there is always the danger that whistle signals from adjacent units may cause confusion. Battlefield noises will reduce the whistle's effectiveness.

3203. Special

Special signals consist of all special methods and devices used to transmit commands or information. The squad leader, operating at night, may use taps on his helmet or rifle butt to signal *halt, danger, move forward,* or *assemble here*. These signals must be understood and rehearsed prior to their use. Various pyrotechnics and smoke signals may be used as signals to attack, withdraw, mark front lines, indicate targets, and cease or shift fire. Before leaders devise others, they should check with their platoon commander to make sure that they are not using a signal which already has a set meaning.

3204. Arm-and-Hand

a. Signals Used With Combat Formations. Explanation and diagrams of standard arm-and-hand signals used with combat formations are given in figure 3-13. See FM 21-60, *Visual Signals*, for detailed information concerning arm-and-hand signals.

1. **DECREASE SPEED.** Extend the arm horizontally sideward, palm to the front, and wave arm downward several times, keeping the arm straight. Arm does not move above the horizontal.

2. **CHANGE DIRECTION; OR COLUMN (RIGHT OR LEFT).** Raise the hand that is on the side toward the new direction across the body, palm to the front; then swing the arm in a horizontal arc, extending arm and hand to point in the new direction.

3. **ENEMY IN SIGHT.** Hold the rifle horizontally, with the stock in the shoulder, the muzzle pointing in the direction of the enemy. Aim in on the enemy target and be ready to engage him if he detects your presence.

4. **RANGE.** Extend the arm fully toward the leader or men for whom the signal is intended with fist closed. Open the fist exposing one finger for each 100 meters of range.

5. **COMMENCE FIRING.** Extend the arm in front of the body, hip high, palm down, and move it through a wide horizontal arc several times.

6. **FIRE FASTER.** Execute rapidly the signal **COMMENCE FIRING.** For machine guns, a change to the next higher rate of fire is prescribed.

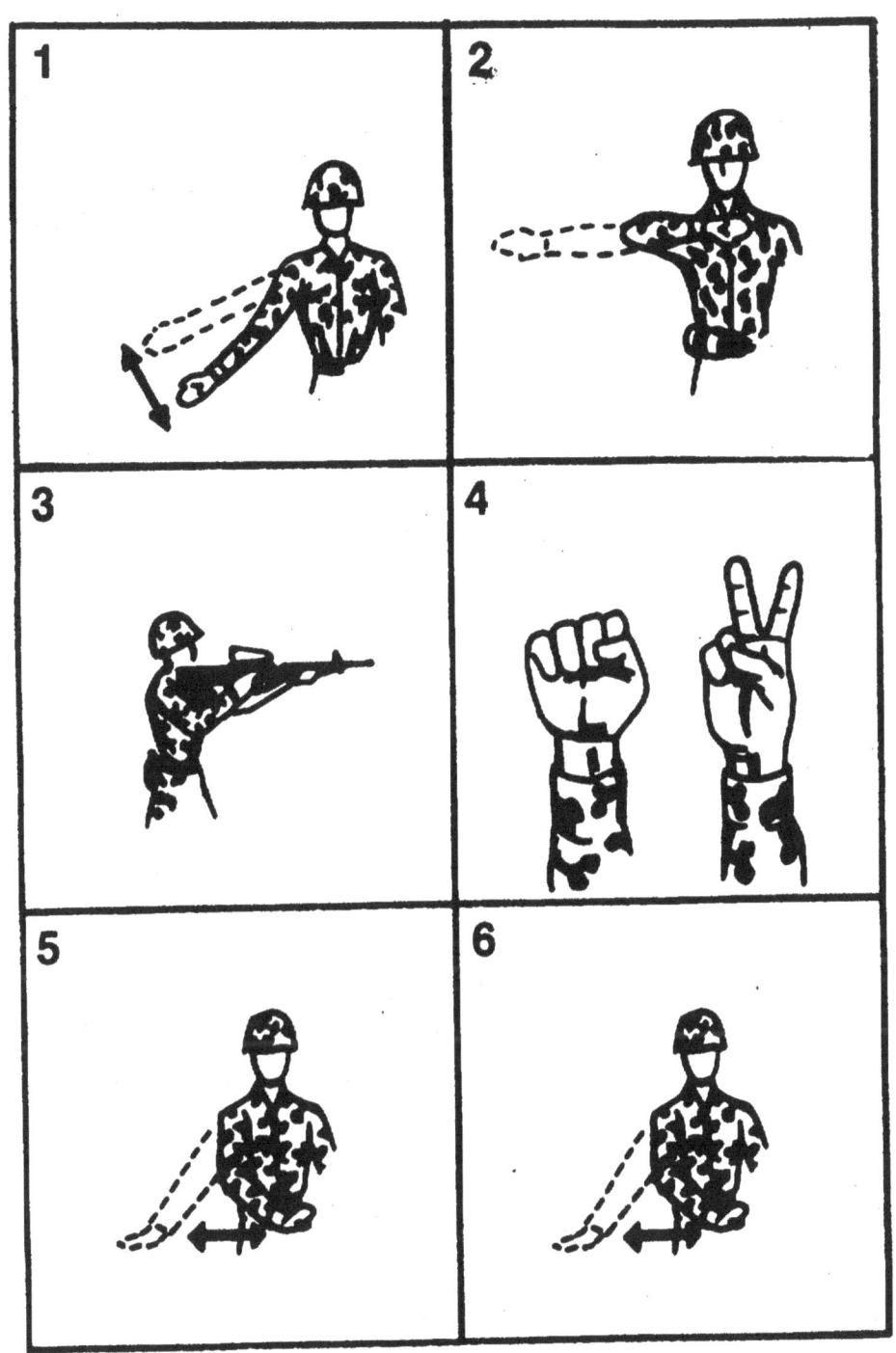

Figure 3-13. Arm-and-Hand Signals.

7. FIRE SLOWER. Execute slowly the signal **COMMENCE FIRING**. For machine guns, a change to the next lower rate of fire is required.

8. CEASE FIRING. Raise the hand in front of the forehead, palm to the front, and swing the hand and forearm up and down several times in front of the face.

9. ASSEMBLE. Raise the hand vertically to the full extent of the arm, fingers extended and joined, palm to the front, and wave in large horizontal circles with the arm and hand.

10. FORM COLUMN. Raise either arm to the vertical position. Drop the arm to the rear, describing complete circles in a vertical plane parallel to the body. The signal may be used to indicate either a troop or vehicular column.

11. ARE YOU READY? Extend the arm toward the leader for whom the signal is intended, hand raised, fingers extended and joined, then raise the arm slightly above horizontal, palm facing outward.

12. I AM READY. Execute the signal **ARE YOU READY**.

13. ATTENTION. Extend the arm sideways, slightly above horizontal, palm to the front; wave toward and over the head several times.

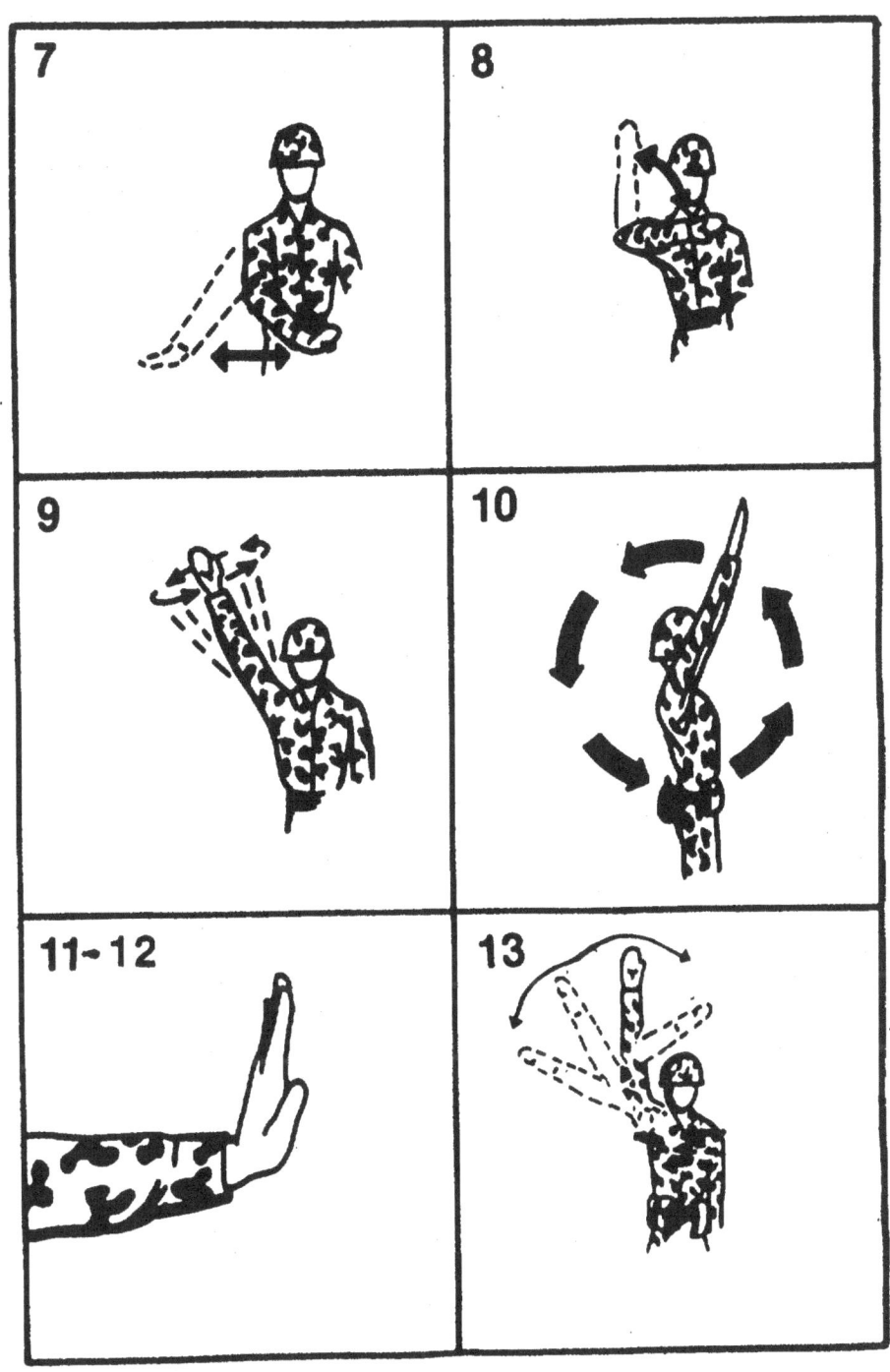

Figure 3-13. Arm-and-Hand Signals. (Continued)

14. SHIFT. Point to individuals or units concerned; beat on chest simultaneously with both fists; then point to location you desire them to move to.

15. ECHELON RIGHT (LEFT). The leader may give this signal either facing towards or away from the unit. Extend one arm 45 degrees below the horizontal, palms to the front. The lower arm indicates the direction of echelon. (Example: for echelon right, if the leader is facing in the direction of forward movement the right arm is lowered; if the leader is facing the unit, the left is lowered.) Supplementary commands may be given to ensure prompt and proper execution.

16. SKIRMISHERS (FIRE TEAM), LINE FORMATION (SQUAD). Raise both arms lateral until horizontal, arms and hands extended, palms down. If it is necessary to indicate a direction, move in the desired direction at the same time. When signaling for fire team skirmishers, indicate skirmishers right or left by moving the appropriate hand up and down. The appropriate hand does not depend on the direction the signaler is facing. Skirmishers left will always be indicated by moving the left hand up and down; skirmishers right, the right hand.

17. WEDGE. Extend both arms downward and to the side at an angle of 45 degrees below the horizontal, palms to the front.

18. VEE. Extend arms at an angle of 45 degrees above the horizontal forming the letter V with arms and torso.

19. FIRE TEAM. The right arm should be placed diagonally across the chest.

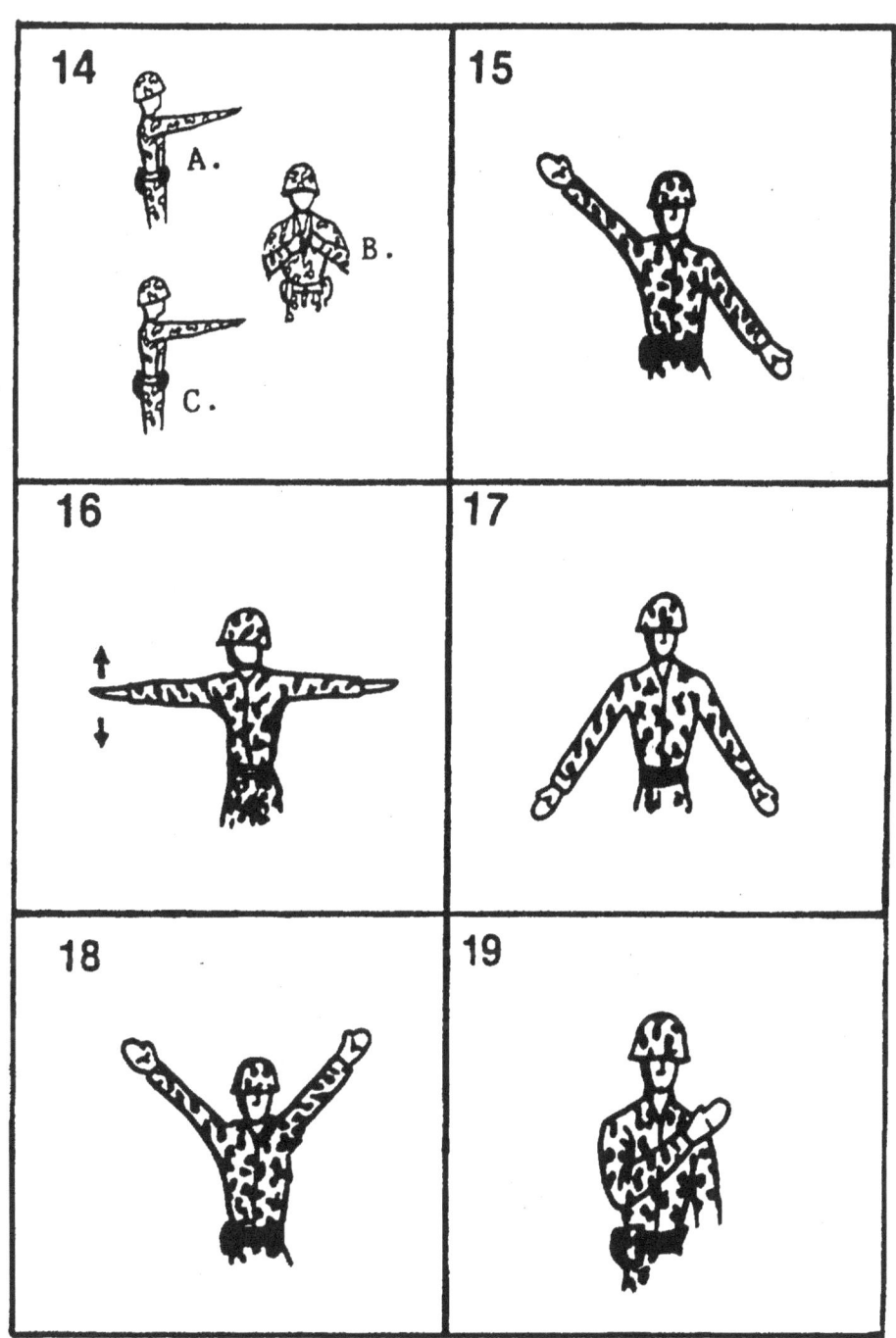

Figure 3-13. Arm-and-Hand Signals. (Continued)

20. SQUAD. Extend the hand and arm toward the squad leader, palm of the hand down; distinctly move the hand up and down several times from the wrist, holding the arm steady.

21. PLATOON. Extend both arms forward, palms of the hands down, toward the leader(s) or unit(s) for whom the signal is intended and describe large vertical circles with hands.

22. CLOSE UP. Start signal with both arms extended sideward, palms forward, and bring palms together in front of the body momentarily. When repetition of this signal is necessary, the arms are returned to the starting position by movement along the front of the body.

23. OPEN UP, EXTEND. Start signal with arms extended in front of the body, palms together, and bring arms to the horizontal position at the sides, palms forward. When repetition of this signal is necessary, the arms are returned along the front of the body to the starting position and the signal is repeated until understood.

24. DISPERSE. Extend either arm vertically overhead; wave the hand and arm to the front, left, right, and rear, the palm toward the direction of each movement.

25. LEADERS JOIN ME. Extend arm toward the leaders and beckon leaders with finger as shown.

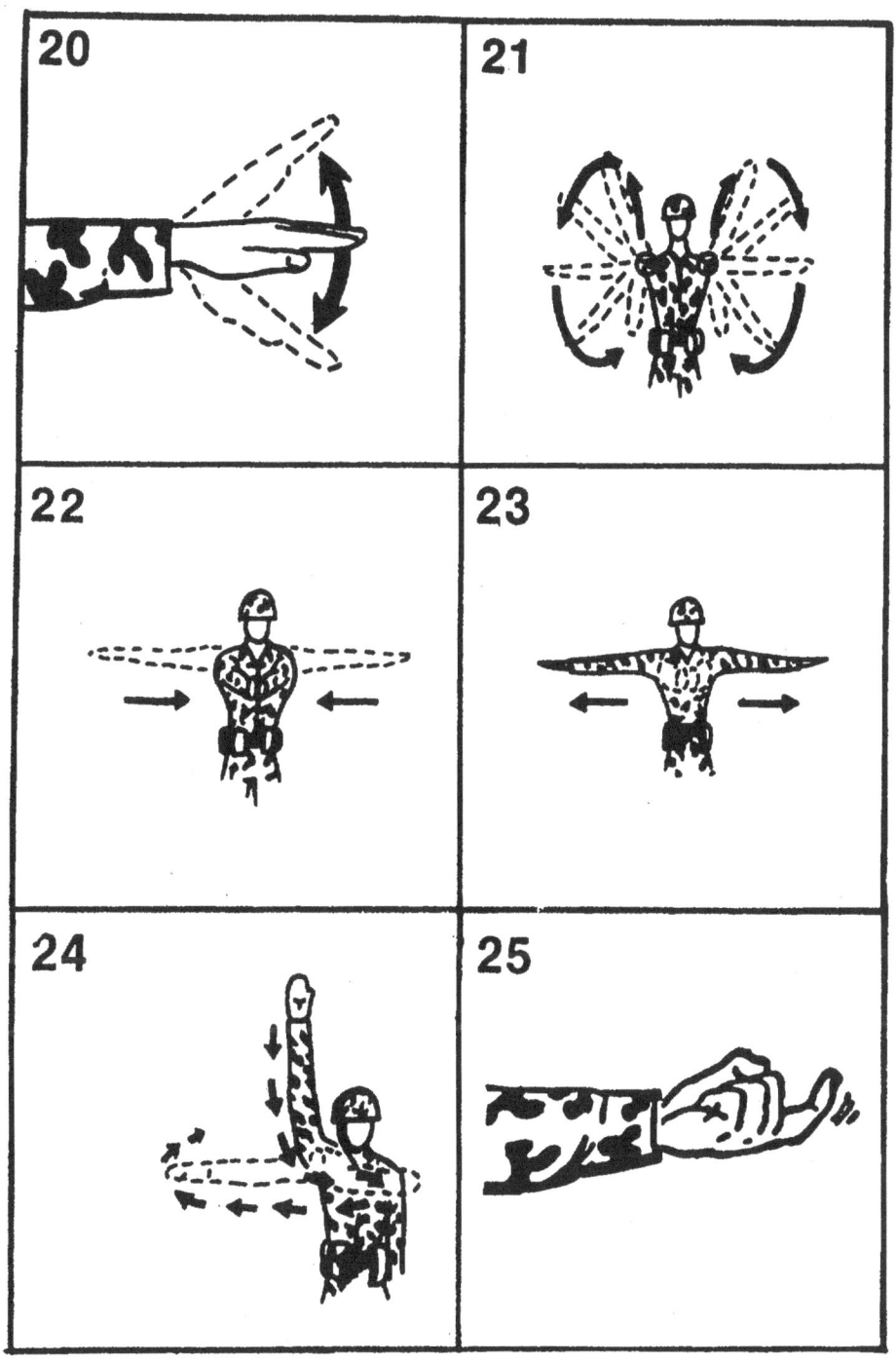

Figure 3-13. Arm-and-Hand Signals. (Continued)

26. I DO NOT UNDERSTAND. Face toward source of signal; raise both arms sidewards to the horizontal at hip level, bend both arms at elbows, palms up, and shrug shoulders in the manner of the universal *I don't know*.

27. FORWARD, ADVANCE, TO THE RIGHT (LEFT), TO THE REAR (USED WHEN STARTING FROM A HALT). Face and move in the desired direction of march; at the same time extend the arm horizontally to the rear; then swing it overhead and forward in the direction of movement until it is horizontal, palm down.

28. HALT. Carry the hand to the shoulder, palm to the front; then thrust the hand upward vertically to the full extent of the arm and hold it in that position until the signal is understood.

29. FREEZE. Make the signal for **HALT** and make a fist with the hand.

30. DISMOUNT, DOWN, TAKE CÓVER. Extend arm sideward at an angle of 45 degrees above horizontal, palm down, and lower it to side. Both arms may be used in giving this signal. Repeat until understood.

31. MOUNT. With the hand extended downward at the side with the palm out, raise arm sideward and upward to an angle of 45 degrees above the horizontal. Repeat until understood.

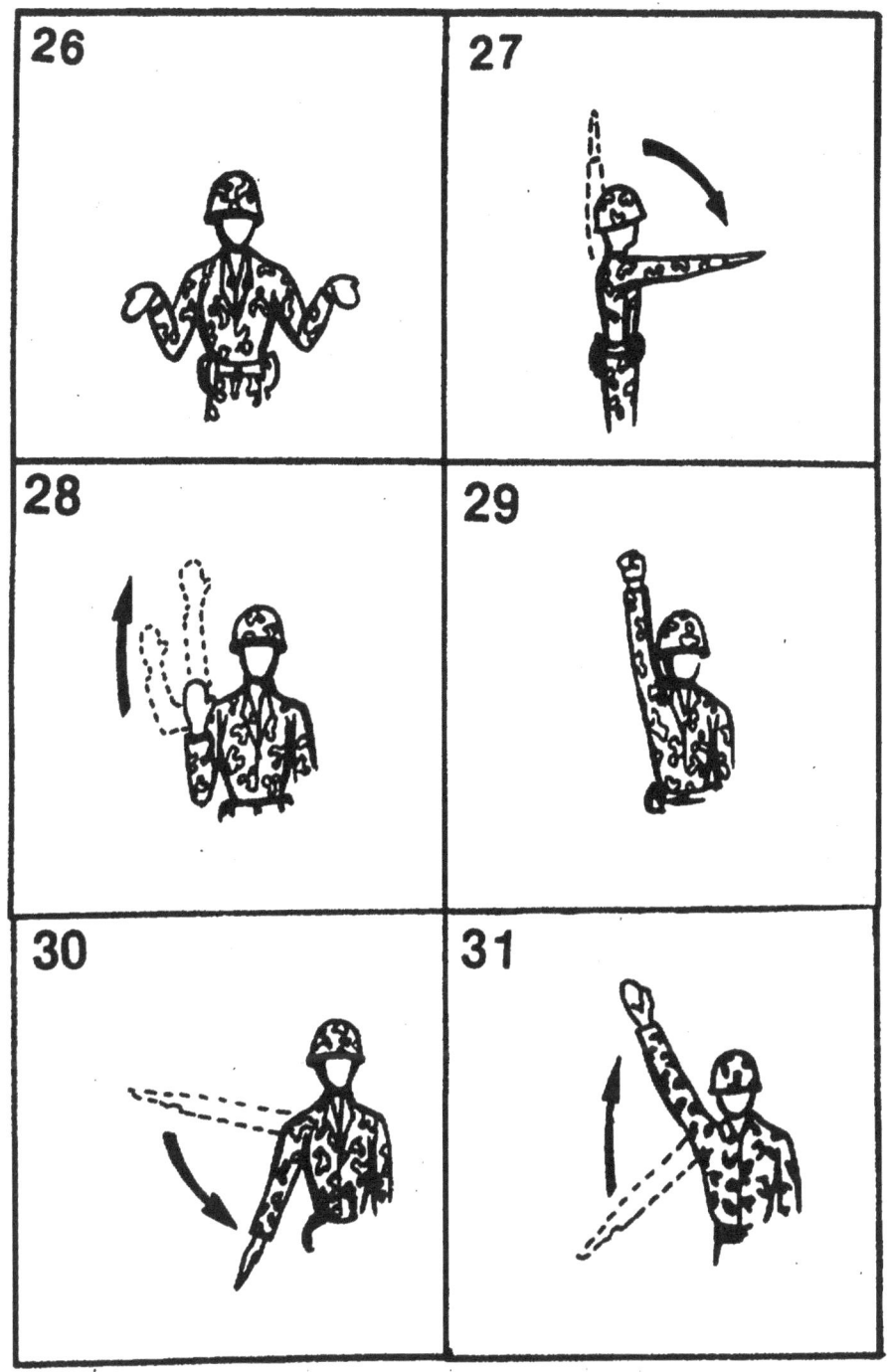

Figure 3-13. Arm-and-Hand Signals. (Continued)

32. DISREGARD PREVIOUS COMMAND; AS YOU WERE. Face the unit or individual being signaled, then raise both arms and cross them over the head, palms to the front.

33. RIGHT (LEFT) FLANK (VEHICLES, CRAFT, OR INDIVIDUALS TURN SIMULTANEOUSLY). Extend both arms in direction of desired movement.

34. INCREASE SPEED, DOUBLE TIME. Carry the hand to the shoulder, fist closed; rapidly thrust the fist upward vertically to the full extent of the arm and back to the shoulder several times. This signal is also used to increase gait or speed.

35. HASTY AMBUSH RIGHT (LEFT). Raise fist to shoulder level and thrust it several times in the desired direction.

36. RALLY POINT. Touch the belt buckle with one hand and then point to the ground.

37. OBJECTIVE RALLY POINT. Touch the belt buckle with one hand, point to the ground, and make a circular motion with the hand.

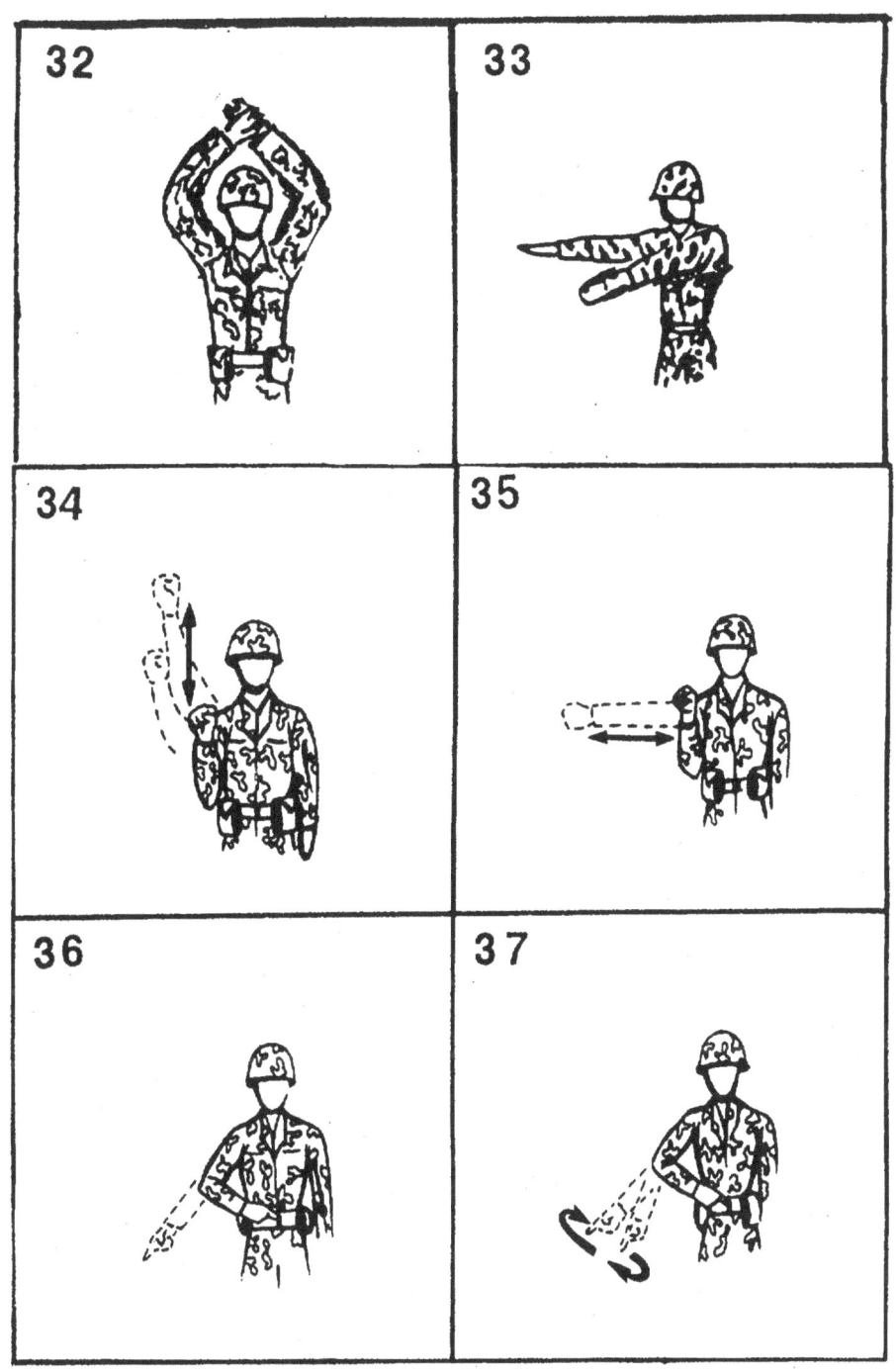

Figure 3-13. Arm-and-Hand Signals. (Continued)

b. Signals Used With Helicopter Operations. Explanations and diagrams of standard arm-and-hand signals used during helicopter operations are given in figure 3-14.

1. **TO PREPARE FOR HELICOPTER GUIDANCE.** Extend arms above the head, palms facing inboard.

2. **TO DIRECT THE HELICOPTER FORWARD.** Extend the arms and hands above the head, palms facing away from the helicopter. Move the hands in such a motion as to simulate a pulling motion.

3. **TO DIRECT THE HELICOPTER BACKWARD.** Extend the arms and hands, palms to the waist, facing the helicopter. Move the hands to simulate a pushing motion.

4. **TO DIRECT THE HELICOPTER TO EITHER SIDE.** Extend one arm horizontally sideways in direction of movement and swing other arm over the head in the same direction.

5. **TO DIRECT THE HELICOPTER TO LAND.** Cross and extend arms downward in front of the body.

6. **TO DIRECT THE HELICOPTER TO TAKE OFF.** Circular motion of right hand overhead, ending in a throwing motion toward the direction of takeoff.

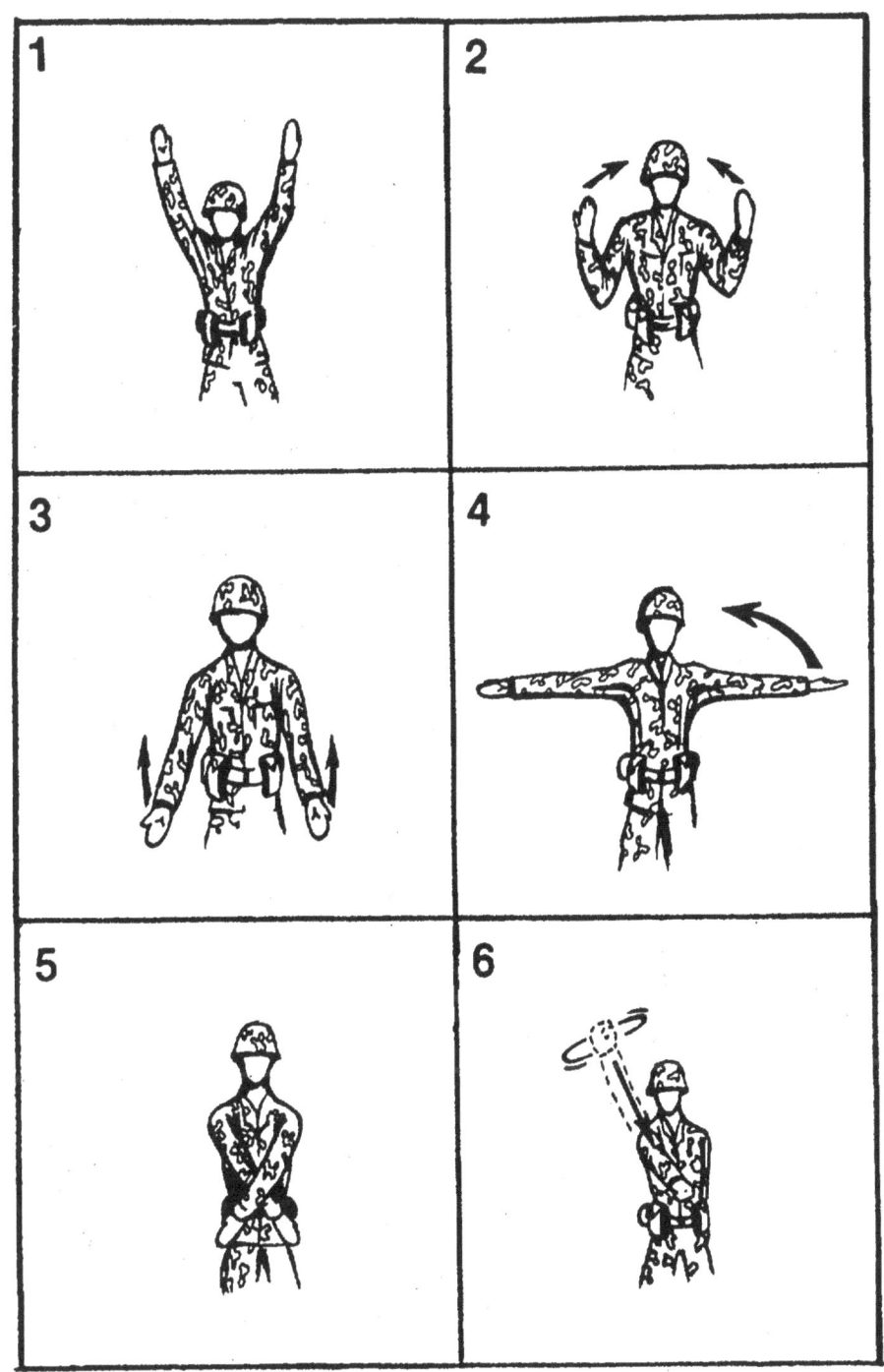

Figure 3-14. Arm-and-Hand Signals.

7. TO DIRECT THE HELICOPTER TO HOVER. Extend arms horizontally sideways, palms downward.

8. TO DIRECT THE HELICOPTER TO WAVE OFF. Arms rapidly waved and crossed over the head.

9. TO DIRECT THE HELICOPTER TO HOOK UP EXTERNAL LOAD. Place the fists in front of body, left fist over the right fist in a *rope climbing* action.

10. TO DIRECT THE HELICOPTER TO RELEASE EXTERNAL LOAD. Left arm extended forward horizontally, fist clenched, right hand making horizontal sliding motion below the left fist, palm downward.

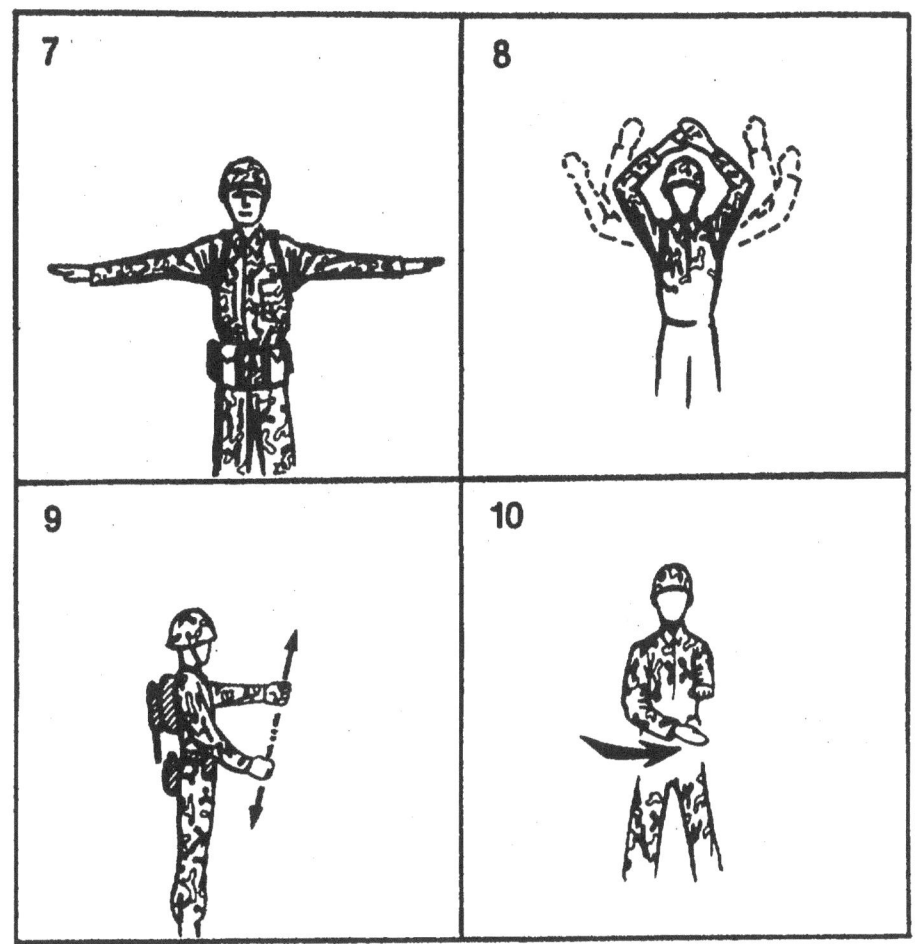

Figure 3-14. Arm-and-Hand Signals. (Continued)

3-51

Chapter 4

Offensive Combat

Section I. General

4101. Purpose

The purpose of offensive combat is to destroy the enemy or his will to fight.

4102. Phases of Offensive Combat

The offensive mission of the squad is to attack. Offensive combat is divided into three phases: preparation, conduct, and exploitation. Each phase is subdivided according to the mission and/or unit involved. In planning and execution, some of the phases may be shortened, omitted, or repeated. Phases and corresponding steps pertinent to the squad are as follows:

a. Preparation
- Movement to the assembly area.
- Reconnaissance and rehearsals.
- Movement to the line of departure.

b. Conduct
- Movement forward of the line of departure to the assault position.
- Advance by fire and maneuver.
- Arrival at the assault position.
- Assault and advance through the assigned objective.
- Consolidation and reorganization.

c. Exploitation
- Continuation of the attack.
- Pursuit.

Section II. Preparation Phase

4201. General

The preparation phase begins with the receipt of the warning order. It ends when the lead element crosses the line of departure or when contact is made with the enemy—whichever comes first. It is usually accomplished in three steps: movement to the assembly area, final preparations in the assembly area, and movement to the line of departure.

4202. Movement to the Assembly Area

The disposition of the squad during the movement to the assembly area is influenced by the size and proximity of the enemy, as well as the squad's location in the column. The route column, tactical column, and approach march (see par. 4204a) are troop formations used in the movement to the assembly area.

a. Route Column. When probability of *contact* with the enemy is *remote* the movement is made in route column. Units within the column are administratively grouped for ease of control and speed of movement. Commanders normally march at the head of their units. This formation may be called an administrative column.

b. Tactical Column. The tactical column is adopted when the enemy situation changes from *contact remote* to *contact possible*. Units within the column are grouped to permit prompt movement into combat formations. The rifle squad may be used as:

(1) **Part of the Main Body.** When the squad marches as part of the main body, the squad leader's primary duties involve the supervision of march discipline within his squad.

(2) **Connecting Elements.** Connecting elements are files or groups which are used to maintain contact between the units of the command. Connecting files are individuals who are sent out to maintain contact between units. A connecting group consists of one or more fire teams. They may be classified as either flank or column connecting files or groups, depending upon their mission. (See fig. 4-1.) The use of connecting files or groups is governed primarily by visibility.

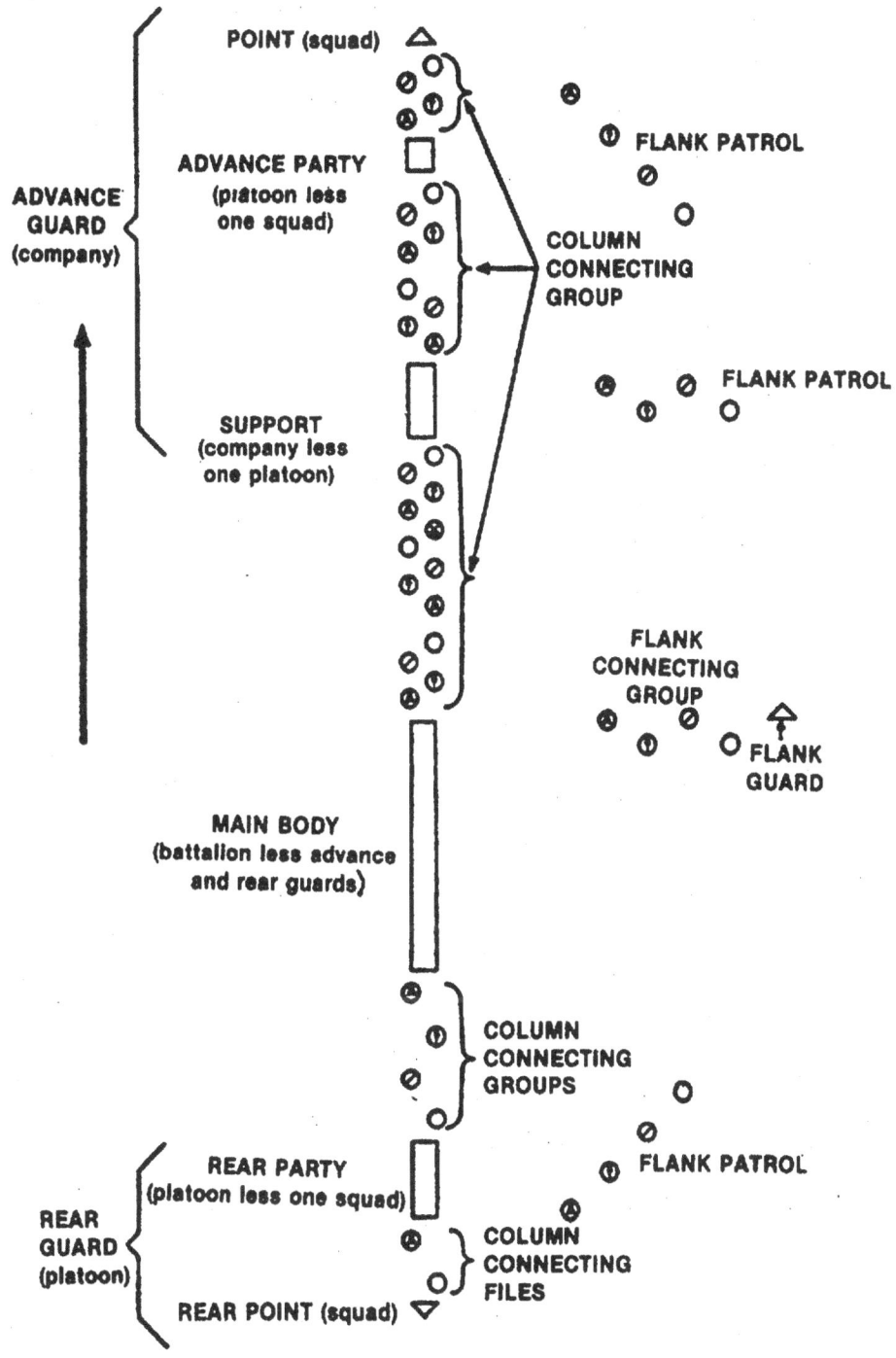

Figure 4-1. Connecting Elements in a Tactical Movement.

(a) Flank Connecting Files or Groups. Connecting files or groups that maintain contact with units, guards, or patrols on the flanks are called flank connecting files or groups. The primary mission of the connecting files or groups is to report the location and situation of the unit on the flank with which they are maintaining contact. They may also cover any gaps which exist between the units, giving warning of and resisting any hostile penetration.

(b) Column Connecting Files or Groups. Individuals or fire teams used to maintain visual contact between elements in the tactical column are called column connecting files or groups. Contact between the point and the advance party is provided by either connecting files or a connecting group consisting of one fire team. Between larger units of the advance guard, main body, and rear guard, a connecting group may consist of a squad.

(3) Point of Advance Guard

(a) The point precedes the advance party along the axis of advance (the general direction of movement for a unit). The distance between the point and advance party is prescribed by the commander of the advance party. Its mission is to prevent an enemy in the immediate vicinity of the route of march from surprising the following troops, and to prevent any undue delay of the column. Possible ambush sites such as stream crossings, road junctions, small villages, and defiles are thoroughly probed by the point.

(b) Formations for the point are prescribed by the squad leader. Generally, the squad uses a wedge or open column formation depending on the terrain. (See fig. 4-2.) When the squad is advancing in the wedge formation, the leading fire team moves on the edges of the road or trail. The two fire teams in the rear march off the road or trail, one team on either side of it. When the road or trail is bound by thick vegetation, or there is a need for haste, the formation of the point is usually a squad column. The fire teams may also be in column formation and advance along alternate sides of the road or trail. In any case, the formation for the point is prescribed by the squad leader, and it is his responsibility to change the formation when the need arises.

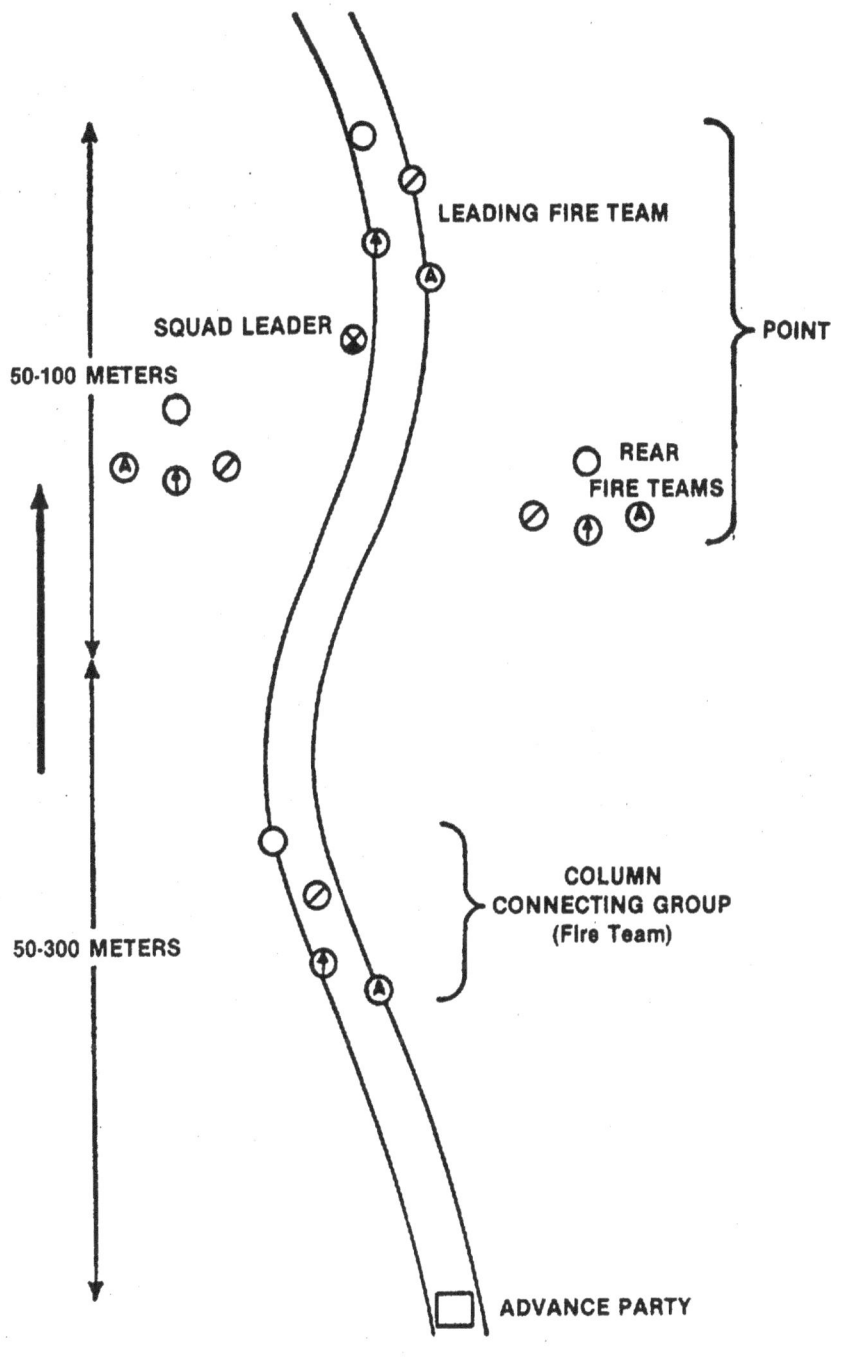

Figure 4-2. The Point in Open Terrain.

4-5

(c) The squad leader assigns each fire team a sector of observation and the fire team leaders assign each individual a sector of observation. Individual sectors of observation should overlap, so there are no gaps in the squad or fire team sectors of observation. This ensures the all-round observation essential for the proper security of the point. (See figs. 4-3 and 4-4.)

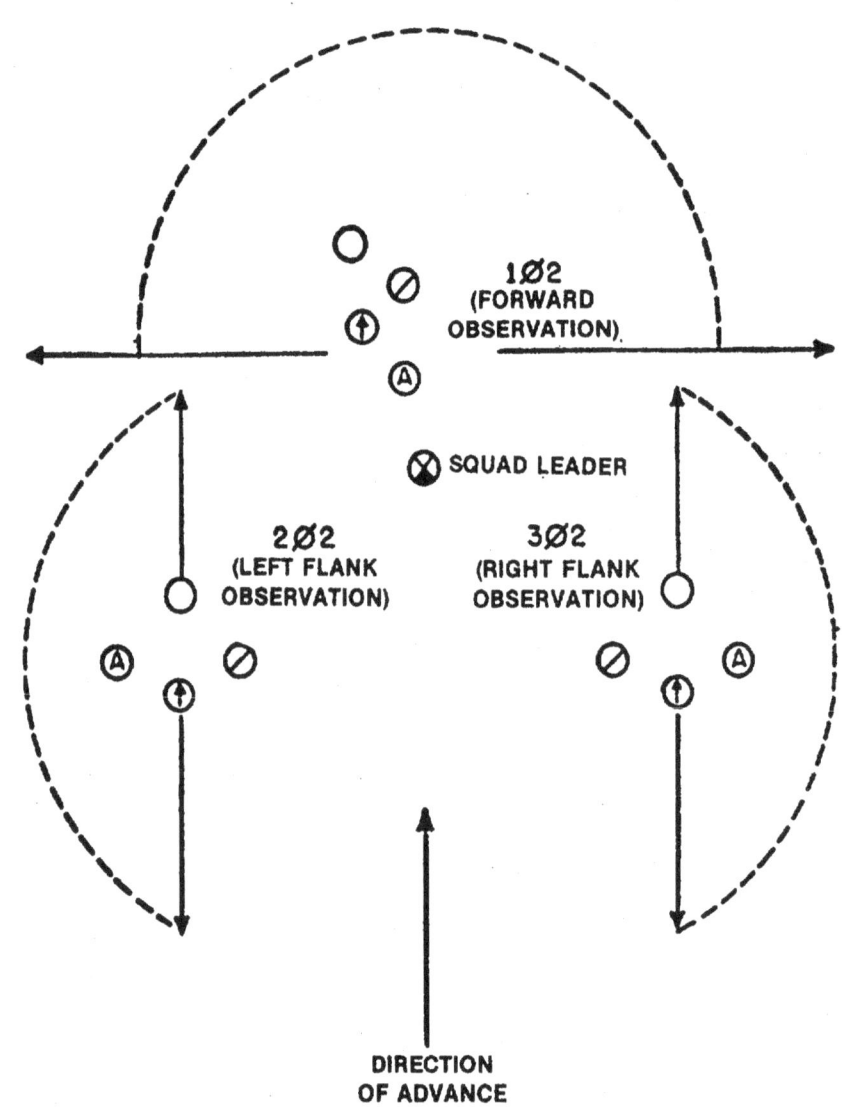

Figure 4-3. Fire Team Sectors of Observation.

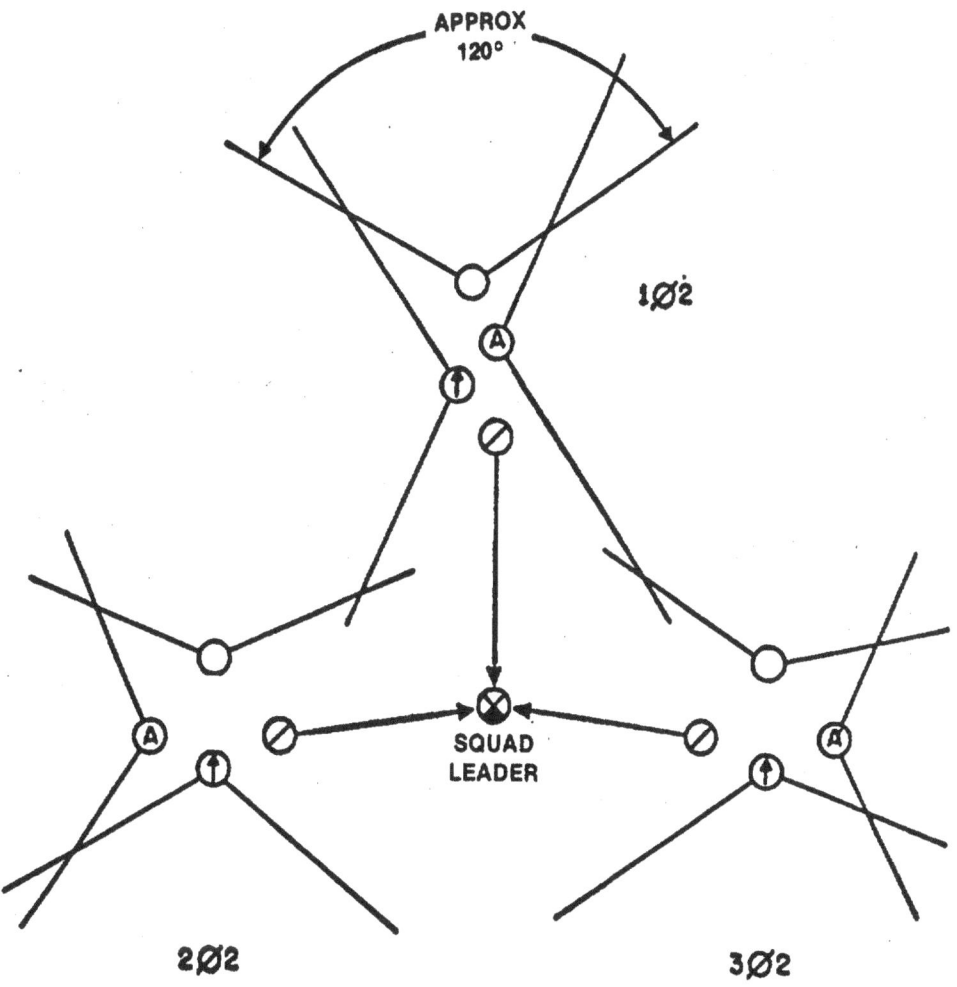

Figure 4-4. Individual Sectors of Observation.

(d) The squad leader of the point generally places himself just to the rear of the leading fire team. From this position, he can most effectively control his squad. He is far enough to the rear to avoid being pinned down by the initial burst of any enemy fire, and yet far enough forward for continuous reconnaissance which enables him to make his estimate of the situation and decision in a minimum of time. The squad leader of the point and the fire team leaders must continually check to see that all members of the squad are alert and vigilant at all times. Weapons are carried ready for instant use. Whenever possible, the point uses arm-and-hand signals for communication.

(e) The point engages all enemy elements within effective range. The squad leader reports contacts to the advance party commander and informs him of the enemy situation and the action he is taking. If the enemy resistance is weak in comparison to the strength of the point, the squad leader initiates a plan to close immediately with the enemy and destroy him. If the enemy resistance is greater than the strength of the point, the squad attacks in a manner that forces the enemy to open fire and disclose his disposition and strength. Such aggressive action materially assists the advance party commander in arriving at a correct estimate of the situation. When the point makes visual contact with an enemy along the route of march but beyond effective range, the advance party commander is notified and the advance continues until contact is made with the enemy. When the enemy is observed beyond effective range to a flank, the point does not proceed to make contact with the enemy, but instead notifies the advance party commander.

(4) Rear Point. In the same manner that the advance party dispatches a point forward, the rear party employs a point to cover its rear. The formation of the squad serving as the rear point is similar to that of the point of the advance guard, but in reverse order. The squad generally employs a vee or a column formation. The squad leader positions himself at the head of the rearmost fire team. This formation is easy to control, provides all-round security, favors fire and maneuver to the flanks, and the fire is adequate in all directions. (See fig. 4-5.) The rear point stops to fire only when enemy action threatens to interfere with the march. Any observed enemy activity is reported to the rear party commander. The rear point cannot expect reinforcement by other troops. It repels all enemy attacks vigorously. If the enemy threatens to overrun the rear point, a covering force from the rear party takes up a position to cover the rear point. When forced back, the rear point withdraws around a flank or along a designated route so as not to mask the fire of the covering force. (See fig. 4-6.)

(5) Point of Flank Guard. The missions, actions, and formations of a squad when serving as the point of a flank guard are the same as when the squad is acting as the point of an advance guard.

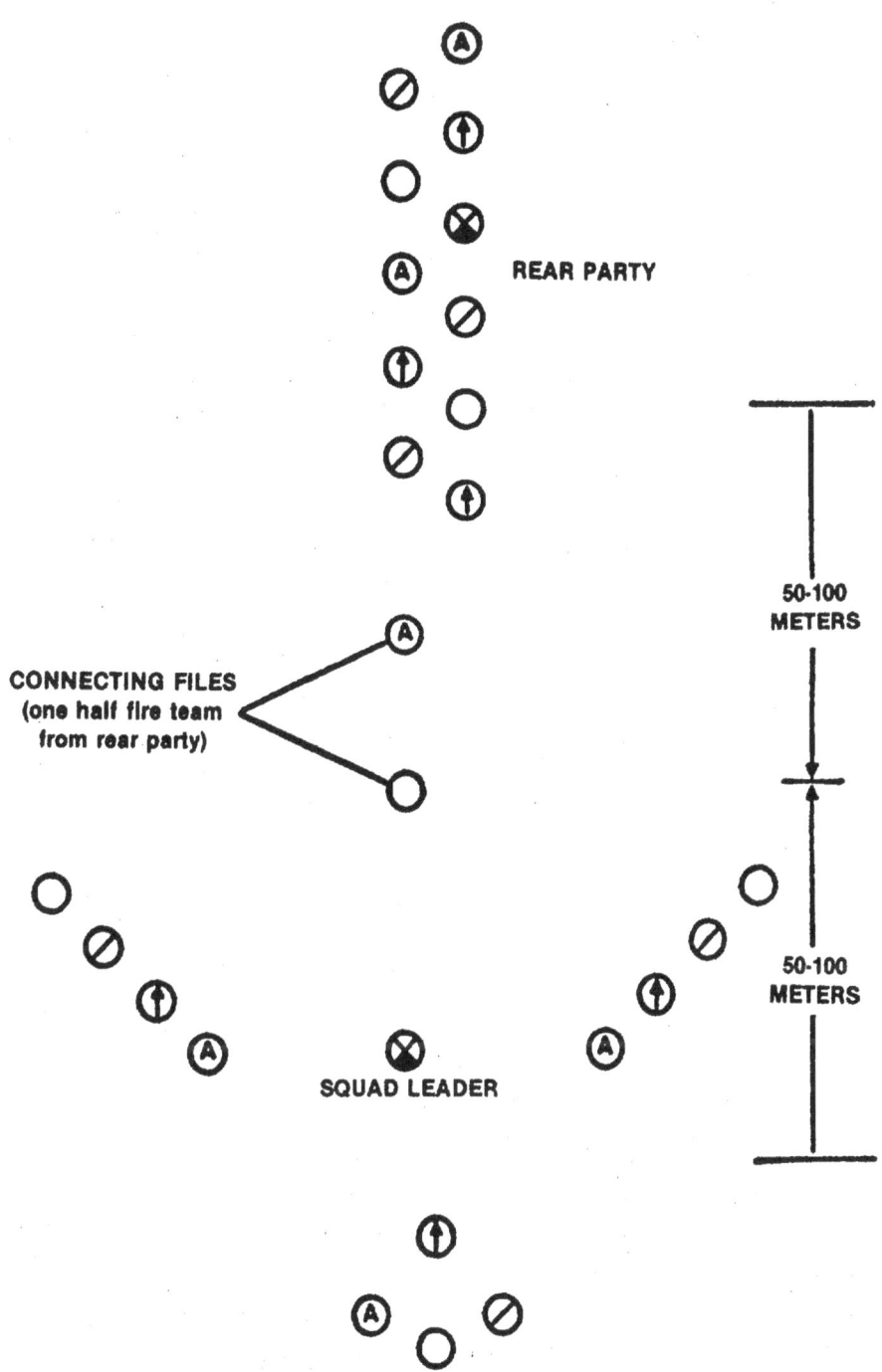

Figure 4-5. The Squad as Rear Point.

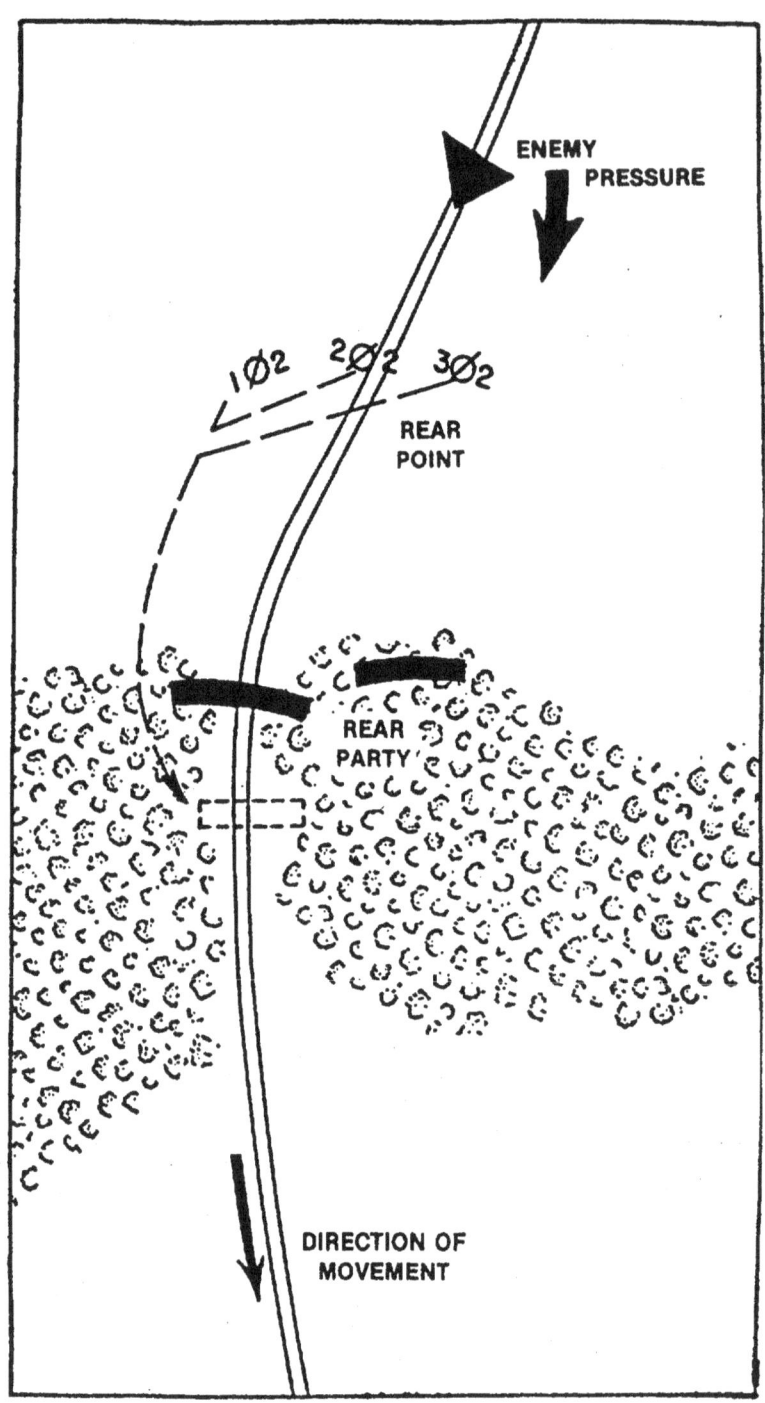

Figure 4-6. Withdrawal of the Rear Point.

(6) Flank Patrol

(a) Rifle squads are often detailed as flank security patrols. A flank patrol may be ordered to move to and occupy an important terrain feature on the flank of the advance, or to move parallel to the column at a prescribed distance from it, the distance depending on the speed of the column and the terrain. When vehicles or helicopters are available and terrain permits their use, it is highly desirable to provide the patrol with transportation.

(b) When moving on foot parallel to the column, the patrol adopts formations based upon considerations of terrain, speed, and self-protection. In open terrain, a wedge formation is usually the best. In heavily wooded terrain, the patrol might use the squad column. The leading fire team serves as the scouting element of the patrol.

(c) The patrol moves so as to prevent the enemy from placing effective small arms fire on the column. It investigates areas likely to conceal enemy elements or provide them good observation. The patrol observes from commanding ground and moves rapidly from point to point, keeping between the protected unit and possible enemy locations.

(d) Enemy patrols moving away from the main body are reported, but are not fired upon unless otherwise directed. All other hostile forces within effective range are engaged immediately by the patrol. If the enemy opens fire on either the patrol or the column, the patrol determines the strength and dispositions and reports this information promptly to the unit or column commander. The patrol resists any enemy attack until ordered to withdraw.

(7) Security for the Halted Column; March Outpost

(a) A march outpost is the outpost which is always established by a marching unit making any temporary halt. It is established by the advance, flank, and rear guards who occupy critical terrain features controlling the approaches to the halted column. Special attention is given to the flanks.

(b) The mission of the march outpost is to protect the halted column from surprise attack by the enemy. If attacked, the march outpost engages the enemy, thus allowing the column time to take up a position from which to repel the attack.

(c) The squad is often detailed as an element in a march outpost. When the squad is so detailed, the platoon commander informs the squad leader of the situation, the outpost position to be occupied, to whom and where reports of enemy activity are to be sent, and the anticipated duration of the halt.

(d) Upon arriving at the prescribed location and making a hasty reconnaissance, the squad leader positions his fire teams where they can observe and defend all avenues of approach leading into the squad area of responsibility. He ensures alert observation by detailing sentinels in pairs and arranging for frequent reliefs. The squad does not abandon its outpost until it receives explicit orders to rejoin its unit. (See fig. 4-7.)

(8) **Termination of the Tactical Column.** The tactical column normally ends when a unit occupies an assembly area to prepare for the attack. However, the enemy situation may cause a unit to deploy into the approach march from the tactical column without occupying an assembly area.

4203. Final Preparations in the Assembly Area

a. Assembly Area. An assembly area is an area where units assemble prior to further tactical action. An assembly area should provide cover, concealment, and security from ground or air attack; it should be large enough to allow for unit dispersion and have ready access to suitable routes forward. When possible, the assembly area should be located beyond the effective range of hostile flat trajectory weapons. Final preparations for the attack are normally completed when the squad is in the assembly area. Those not completed may be accomplished in the attack position. These preparations include reconnoitering, formulating plans, and issuing orders. Also—

- Additional ammunition is drawn and distributed.
- Weapons, equipment, and personnel are checked for readiness.
- Equipment not required for the attack is collected and staged for later pickup.
- Extra or special equipment needed for the operation is obtained and issued.
- Personnel are allowed to rest as much as possible.
- Communication equipment is checked. Leaders must ensure that the required frequencies and call signs are on hand.

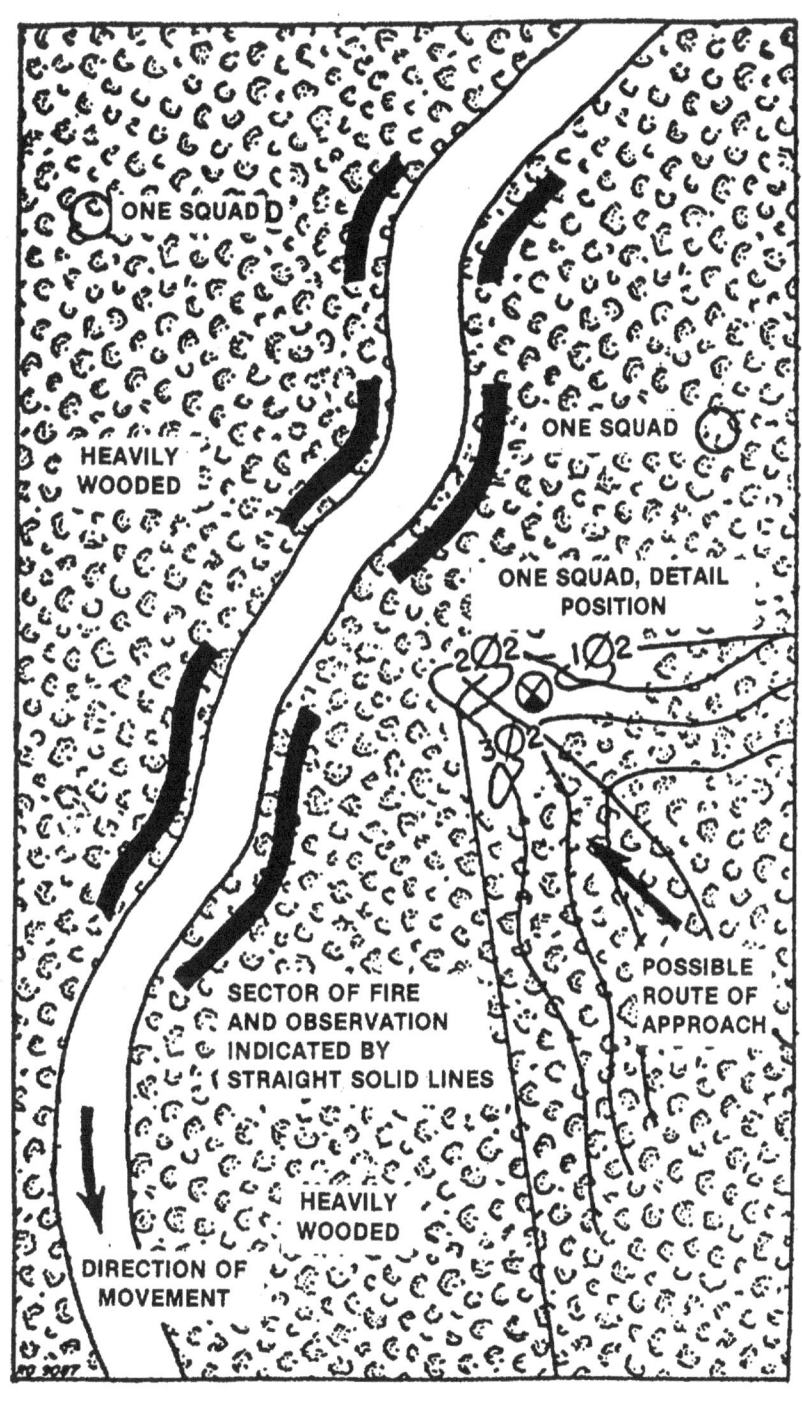

Figure 4-7. The Squad as an Element of a March Outpost.

b. Troop Leading Procedures. The troop leading steps are a useful method for assisting the squad leader in preparing for an attack. Explanation of these procedures is included in appendix C.

c. Attack Plan. The fourth step of the troop leading procedures requires the squad leader to complete his attack plan. There are two methods of attack for a squad: (1) a single envelopment (see par. 4302h[1]), and (2) a frontal attack (see par. 4302h[2]).

d. Issue Attack Order. Having completed his plan of attack, the squad leader issues his attack order. (See appendix E for the five-paragraph order format). An example of the type of information included in a five-paragraph order follows:

Unit: 3d Squad, 2d Platoon with SMAW team attached.

Terrain Orientation (given prior to issuance of the order): "That direction is north (pointing). Notice that ridge line to our front, generally east and west. Note that line of thick low bushes. Notice that finger in the center of the ridge running to the north with the trail up its left side (pointing)."

   (1) Situation

      (a) Enemy Forces. "The main enemy force is defending along that high ground to north (pointing). His automatic weapons are well dug in near the military crest. I observed some riflemen in open fighting holes around the automatic weapons. I can't tell exactly how many of them there are."

      (b) Friendly Forces. "Our platoon attacks at 0815 to seize the left half of the company objective. The 2d squad is on our left. The 1st platoon attacks on our right to seize the right half of the company objective."

      (c) Attachments and Detachments. "A SMAW team will be attached to us. They're ours at 0745."

   (2) Mission. "Our squad attacks at 0815 on a frontage of 125 meters to seize the right ⅓ of the platoon objective."

(3) Execution. (It is here where the commander makes his intent clear.)

(a) Concept of Operation. "We will cross the line of departure in a squad line, fire teams in wedge and attack the objective frontally, guiding on that trail up the left of the finger. Fire teams go to skirmishers at the assault position."

(b) Subordinate Missions. "Cpl Jones, I want your team on the left side of the line. You will be the base fire team for the attack. Guide on that trail going up the finger. Watch for the 2d squad on your left. Take the left one third of the finger from the red dirt slash on the left to the trail on the right."

"Cpl Brown, your team will be in the center; take the middle third of the knob from the trail on the left to the narrow path on the right."

"Cpl Smith, your fire team will be on the right. Watch for the 1st platoon on your right. Take the right half of the knob, from the narrow path to the tree line."

"SMAW team, general support. Follow in trace of Cpl Brown's fire team by about 50 meters. Engage targets of opportunity during our movement. Consolidate on the objective with Cpl Brown's team upon completion of the assault."

(c) Coordinating Instructions. "Line of departure is that line of thick low bushes. Direction of attack is north. The tentative assault position will be the depression approximately 50 meters short of the objective."

(4) Administrative and Logistics. "Cpl Jones, send two men to the platoon sergeant to draw 14 frag hand grenades. Each fire team will get four. I will carry the other two. A corpsman will be directly to our rear. Rations and ammo will be issued on the objective. Evacuate casualties and prisoners of war to the immediate rear."

(5) Command and Signal

(a) Communication and Special Signal Instructions. "The red smoke grenades will be used by the platoon commander to signal supporting fires to lift."

(b) Location of the Commander. "I will be with the center fire team in the assault and when we consolidate on the knob. The lieutenant will be with the 2d squad. It is now 0709. Any questions?"

4204. Movement to the Line of Departure

Upon leaving the assembly area, the squad makes a rapid and continuous advance to the line of departure. If necessary, a brief halt to effect last minute coordination and to assume initial combat formations may be made in the attack position. If the squad is subjected to artillery or mortar fire along the route, it moves quickly through or around the impact area. See figure 4-8 for tactical control measures used by the squad in an attack.

a. Approach March. The squad leaves the assembly area and continues the movement toward the enemy in the approach march formation. The approach march formation is used when enemy *contact* is *imminent*. The column establishes guards to the front, flanks, and rear, as appropriate. Elements within the column may be fully or partially deployed in the attack formation. Generally, the advance is made by bounds, stopping on easily recognizable terrain features to coordinate the advance. During the approach march, the squad and fire teams take maximum advantage of cover and concealment along the route. (See fig. 4-9.)

(1) Initial Formation. Upon assuming an approach march formation, the platoon commander usually prescribes initial squad formations. As the march progresses, however, the squad leaders order formation changes in accordance with the terrain, the frontages assigned, and the likelihood of enemy contact.

(2) Base Squad. A base squad is designated by the platoon commander to assist in maintaining direction, position, and rate of march. Other squads will guide on the base squad.

(3) Duties of the Squad Leader. The squad leader regulates his squad's advance on the base squad, or if his squad is the base squad, he advances it as directed by the platoon commander. As he moves, he studies the ground to the front in order to take advantage of cover and concealment and to control the movement of his fire teams. He also maintains direction and makes minor detours to take advantage of better terrain.

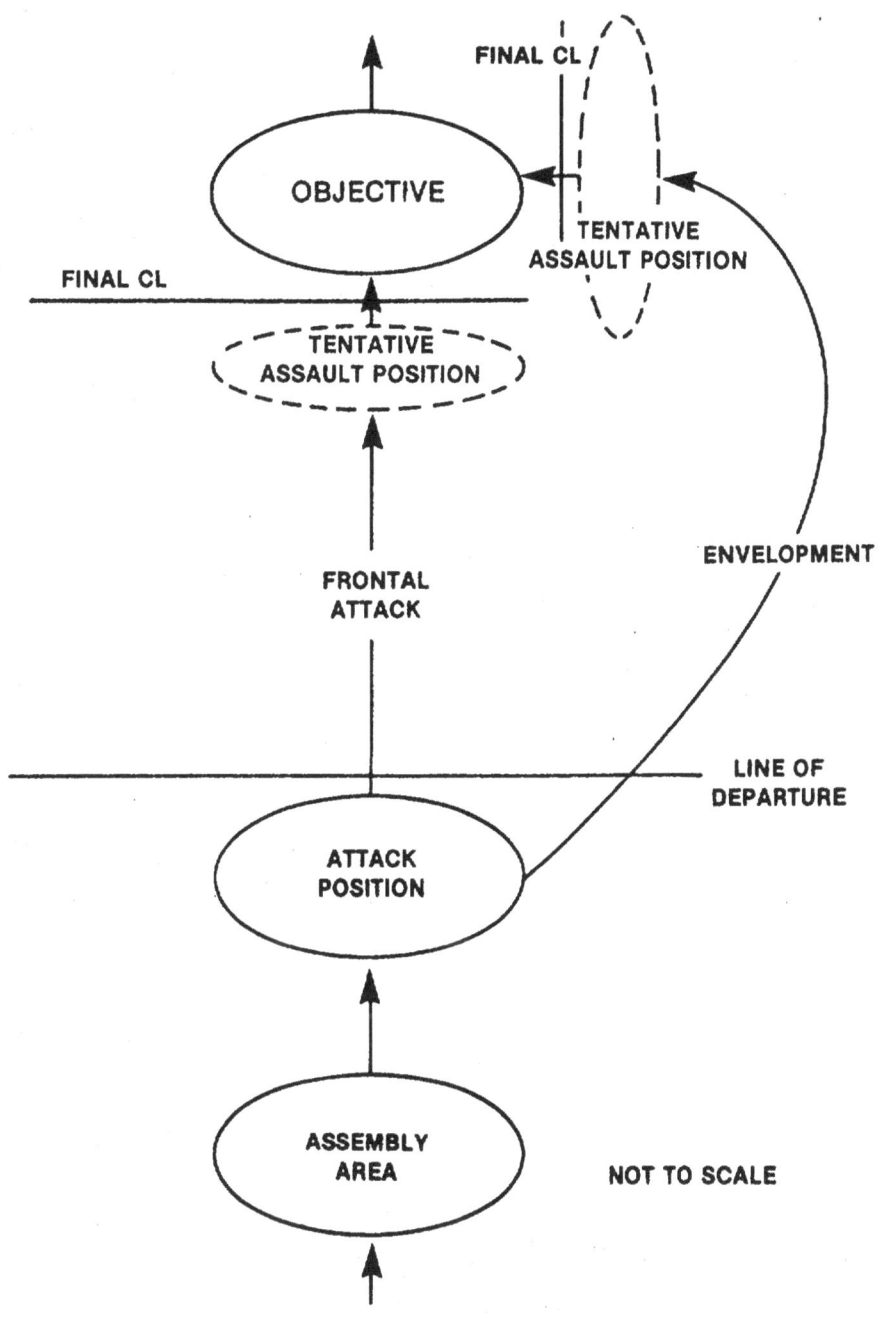

Figure 4-8. Typical Tactical Control Measures Used by Rifle Squad in a Dismounted Daylight Attack (Schematic).

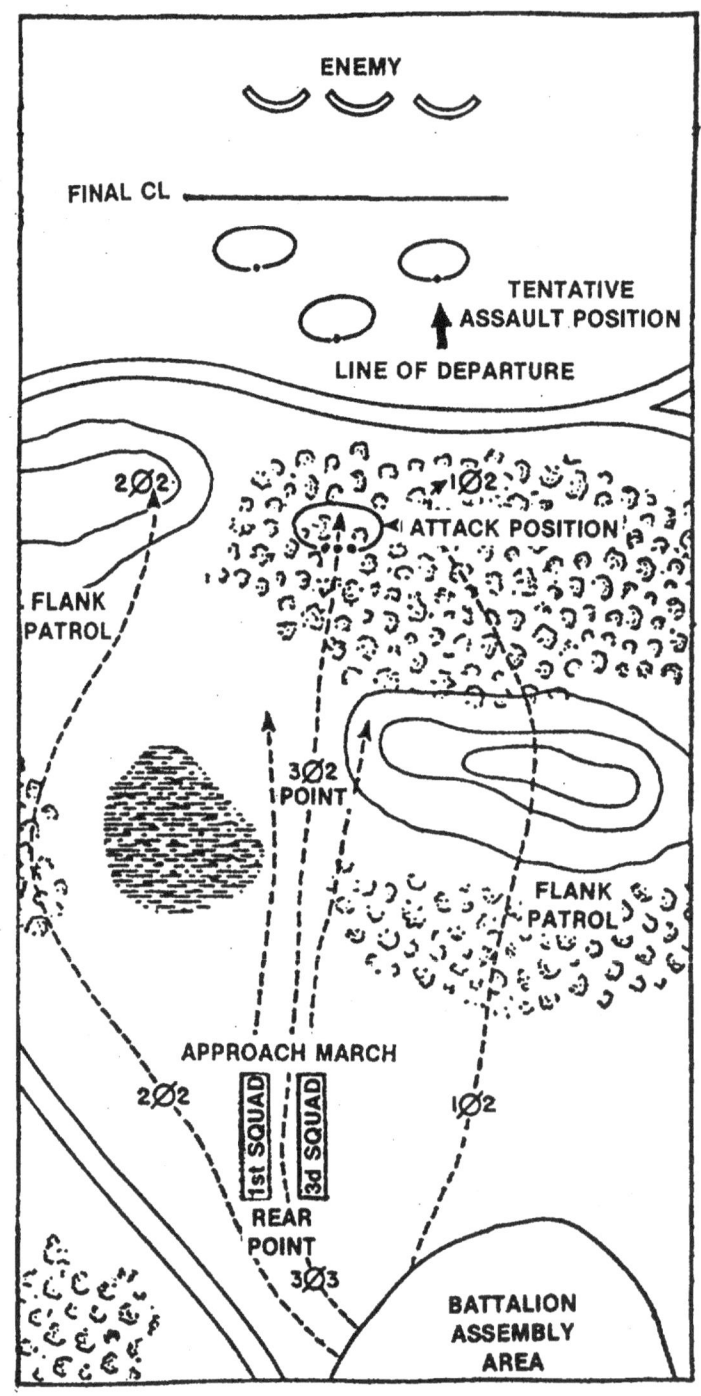

Figure 4-9. Rifle Platoon from Assembly Area to the Assault Position.

(4) Scouting Fire Team

(a) When a rifle platoon in the approach march is not preceded by friendly troops, it uses its own scouting elements. The scouting element is usually one fire team; however, an entire squad may be used. A fire team used as a scouting element is called a scouting fire team and is controlled by the platoon commander, assisted by the squad leader. A squad leader whose squad is providing the scouting fire team normally marches near the platoon commander to assist in the control of the scouting fire team. A scouting fire team moves aggressively to cover the front of the advancing platoon and to force the enemy to disclose his position. Formations generally used by a scouting fire team are the wedge or skirmishers. Normally, a scouting fire team scouts a frontage of 50 to 75 meters. If a wider frontage must be covered, the entire squad must be employed, normally using the vee or wedge formation. Scouting fire teams are covered by the platoon or, when the platoon is masked, the fire team covers its own advance. The fire team leader watches constantly for signals from the platoon commander, remaining in visual contact at all times. The distance that the scouting fire team moves ahead of the platoon varies with the terrain. This distance is normally the limit of visibility. In open terrain, the platoon commander usually directs the scouting fire team to move by bounds along a succession of objectives.

(b) When a scouting fire team is directed to advance over open ground to the edge of a woodline, two members go inside the woodline for 50 to 60 meters while the other men of the fire team cover them. When it is determined that the area near the edge of the woods is clear, the fire team leader signals the platoon commander that it is safe to move forward and then moves the remainder of the fire team into the woods. The scouting fire team then occupies and holds a line 50 to 75 meters within the woods and observes toward the front until the platoon comes up. The scouting fire team leader awaits further word from the platoon commander before moving the team further into the woods. When directed, he moves the team forward until they reach the far edge of the woods. The team holds at the edge of the woods and notifies the platoon commander of the situation. The platoon commander moves the platoon to a position where it can cover the scouting fire team and directs the team leader to move out and continue his scouting mission.

(c) When a scouting fire team is fired upon, the individuals immediately take cover, locate targets, and return fire. The scouting fire team leader then determines:

- Location of enemy (range and reference points).
- Extent of position (location of flanks).
- Types of positions (fighting holes, bunkers, obstacles, etc.).
- Number of enemy.
- Enemy weapons (mortars, tanks, etc.).

(d) The platoon commander contacts the leader of the scouting fire team to obtain as much information as possible. The platoon commander then returns control of the scouting fire team to the squad leader. Usually the platoon commander brings up the remainder of his squads, sets up a base of fire, and assaults the enemy position. Should the enemy position prove too strong for the platoon, the platoon usually remains engaged with the enemy as a base of fire while units of the advance party are committed to clear the enemy resistance.

b. Attack Position. The attack position is the last concealed and covered position occupied by assault echelons before crossing the line of departure. It is the location where final coordination, last-minute preparations, and, if not already accomplished, deployment into initial attack formations are effected. When all preparations for the attack are completed in the assembly area, there should be no delay when passing through the attack position. (See figs. 4-8 and 4-9.)

4205. Special Situation

Paragraph 4102 states that in planning and execution, some of the phases of offensive combat may be shortened, omitted, or repeated. An example of this would be when a unit that is occupying defensive positions is in contact with the enemy and is ordered to conduct an attack. Since the unit is already in contact with the enemy, the advance to contact would be omitted. Due to the danger of enemy observation, direct fire weapons and ability to mass indirect fire weapons (artillery and mortars), it is not always possible to move the unit out of the defensive positions back to an assembly area. In this situation, the unit's present defensive positions become the assembly area, attack position, and line of departure. All of the actions and final preparations, which under normal circumstances would be accomplished in the assembly area and attack position, will be accomplished while the majority of the unit remains in place.

Section III. Conduct Phase

4301. General

The conduct phase of offensive combat begins when one of the following occurs:

- The squad is forced to fire on the enemy in order to advance.
- The leading troops cross the line of departure.

4302. Movement From the Line of Departure to the Assault Position

When the squad leader believes he has reached a point where his squad can no longer advance without sustaining casualties, he orders one or two fire teams to fire on the enemy positions while the remainder of the squad moves forward under the protection of this covering fire. The maneuver used in a particular situation is decided by the squad leader based on his rapid estimate of the situation. When the enemy position is isolated and has exposed flanks, the squad leader attempts to maneuver over a covered and concealed route so as to strike the enemy position in the flank or rear. When this is not possible, a frontal attack requiring fire and maneuver is executed.

a. Fire and Maneuver. Fire and maneuver is the process whereby elements of a unit establish a base of fire to engage the enemy, while another element maneuvers to an advantageous position from which to close with and destroy or capture the enemy. Supporting fires from weapons not organic to the unit may be provided. Supporting fires should be followed closely by the advancing troops of the maneuver unit so that the shock effect of the fire upon the enemy will not be lost.

b. Fire and Movement. Once the maneuver element meets enemy opposition and can no longer advance under the cover of the base of fire, it employs fire and movement to continue its forward movement to a position from which it can assault the enemy position. In a maneuvering squad, fire and movement consists of individuals or fire teams providing covering fire while other individuals or fire teams advance toward the enemy or assault the enemy position.

c. Squad Employment. The squad is normally employed as part of the rifle platoon and will be assigned a mission as a base of fire or as a maneuver element. Thus, operating as part of the platoon, a squad assigned as the maneuver element will execute fire and movement, not fire and maneuver. A squad will be required to fire and maneuver when, for example, given a mission such as point squad, flank patrol, or flank guard during a movement to contact, enemy contact is made. The organization of the rifle squad into three fire teams provides the squad leader with the ability to execute fire and maneuver with one or two fire teams employed as the base of fire and one or two fire teams as the maneuver element.

d. Fire Team. The fire team, as the basic fire unit, is restricted to executing only fire and movement.

e. Base of Fire Element. The base of fire covers the maneuver element's advance toward the enemy position by engaging all known or suspected targets. Upon opening fire, the base of fire seeks to gain fire superiority over the enemy. Fire superiority is gained by subjecting the enemy to fire of such accuracy and volume that the enemy fire ceases or becomes ineffective.

f. Maneuver Element. The mission of the maneuver element is to close with and destroy or capture the enemy. It advances and assaults under covering fire of the base of fire element. The maneuver element uses available cover and concealment to the maximum. Depending upon the terrain and effectiveness of the covering fire, the maneuver element advances by team movement; within the team, by fire and movement, employing rushes, or creeping and crawling as necessary. Regardless of how it moves, the maneuver element must continue to advance. If terrain permits, the maneuver element may be able to move forward under cover and concealment to positions within hand grenade range of the enemy.

g. Control of the Squad

(1) Fire team leaders initiate the action directed by the squad leader. In the attack, fire team leaders act as fighter-leaders, controlling their fire teams primarily by example. Fire team members base their actions on the actions of their fire team leader. Throughout the attack, fire team leaders exercise such positive control as is necessary to ensure that their

fire teams function as directed. The squad leader locates himself where he can best control and influence the action. In controlling the squad when taken under enemy fire, the squad leader takes into account the fact that the battlefield is a very noisy and confusing place. If enemy fire is light he may be able to control his fire team leaders by voice, whistle, or arm-and-hand signals. As the volume of enemy fire increases, this type of control becomes impossible. In this situation the squad leader must rely on the skill and initiative of the fire team leaders to carry out the instructions he previously gave them. To maintain control of the squad under heavy enemy fire, the squad leader positions himself near the fire team leader of the designated base fire team. By regulating the actions of the base fire team leader, the squad leader retains control of the squad. The base fire team leader controls the actions of his fire team; the other fire team leaders base their actions on those of the base fire team. This type of control must be practiced and perfected in training if the squad is to be effective in combat.

(2) The base fire team is used by the squad leader to control the direction, position, and rate of movement of the squad. It is not intended that the other fire teams maintain rigid positions in relation to the base fire team; the base fire team is used as a general guide. If another fire team can move forward more rapidly than the base fire team, it should do so. For instance, if the base fire team is receiving enemy fire, but the terrain in front of another fire team provides cover from enemy fire, the latter team should move rapidly forward to a position where they can deliver fire on the enemy. Covering the base fire team's movement by fire takes pressure off them and permits them to move forward. Once the base fire team comes generally abreast, the other fire teams can then resume fire and movement.

h. Use of Maneuver. Once fire superiority has been gained, the squad continues its advance. Fire superiority is maintained throughout the attack in order to ensure the success of any maneuver. Before advancing any part of his squad, the squad leader should assure himself that there is sufficient fire on the enemy position to render return fire ineffective. Two forms of maneuver for the rifle squad are the single envelopment and the frontal attack using rushes. In a single envelopment, the maneuver element attacks against the flank or immediate rear of the enemy's position. The frontal attack exerts pressure against the enemy's front and drives him off the objective.

(1) Single Envelopment. A squad maneuvering against the enemy's flank is normally covered by a supporting attack conducted by another squad acting as the base of fire. The maneuvering squad moves toward the flank of the enemy so as to place itself in a position to make an assault. The maneuvering squad takes advantage of available cover and concealment, keeping the enemy unaware of its movements until the assault begins. When the maneuvering squad commences the assault, the base of fire shifts to another part of the enemy position or ceases firing entirely. If observation permits, it is desirable to have the base of fire lead the maneuvering squad across the objective by fire. The single envelopment splits the enemy's defensive fires; part focuses on the base of fire (supporting attack) and part on the maneuvering squad (main attack), and allows the maneuvering squad to attack over ground of its own choice.

(2) Frontal Attack. When there is no opportunity for maneuver to either flank of the enemy, the maneuvering squad moves directly to the front. The squad leader orders one fire team to advance under cover of fire of the remainder of the squad. Fire teams advance as rapidly as possible to new firing positions, using the cover and concealment available. When a fire team reaches a new firing position, that fire team opens fire. The part of the squad that was providing the covering fire ceases fire and under cover of this newly established covering fire moves forward, using the available cover and concealment. This process is continued until the squad is in position to assault the enemy. The squad leader moves to successive positions from where he can best maintain effective control of his squad. The frontal attack is the most frequently used form of maneuver by the rifle squad. The frontal attack requires less time and coordination and is easier than the single envelopment. However, the attack moves against the enemy's strength and prepared fires and there is little chance he will be surprised.

i. Method of Advance. When making either a single envelopment or a frontal attack, a rifle squad has three methods by which it may move. The squad may move as a unit in a series of squad rushes, as fire teams in a series of alternating fire team rushes, or the members of the squad may move forward singly by individual rushes. The volume of the enemy's fire will determine which method the squad will use. In all three, the element of speed is necessary.

4303. Movement From the Assault Position Through the Objective

The primary object in advancing the attack by fire and maneuver and/or fire and movement is to get part or all of the attacking unit in position to assault the enemy. The position from which the final assault is launched is called the *assault position*. As the attacking squad closes with the enemy, covering fires delivered by both direct and indirect fire weapons on the enemy position increase in intensity. In order to avoid casualties by friendly fire these *supporting fires are ceased or shifted* just prior to reaching the objective. The control measure where this happens is the *final coordination line* (final CL). Both the assault position and final CL are crucial to the assault.

a. Assault Position. The assault position is tentatively established during the squad leader's planning and reconnaissance. It is the position between the line of departure and the objective, from which the assault on the enemy position is launched. The assault position is located as close as the assaulting element can move by fire and maneuver without sustaining casualties from or masking covering direct (base of fire) or indirect fires (artillery and mortar). The assault position should be easily recognizable on the ground and ideally should offer concealment and cover to the attacking force. Here, the final steps are taken to ensure a coordinated assault, and *only a minimum amount of time* should be spent in this position to preclude the enemy from fixing the assault element in place.

(1) When the squad reaches the assault position the squad leader, fire team leaders, and squad members must *quickly* make final preparations for the assault. Unit leaders issue last minute instructions to their men. Squad members armed with the M-16, to include the fire team leader/grenadier, will insert a full magazine. Riflemen, assistant automatic riflemen, and squad leaders should fix bayonets; fire team leaders should load the M-203 with the type of ammunition directed by the squad leader. Automatic riflemen ensure that their weapons contain sufficient ammunition for the assault. If the 200-round ammunition box is being used, a quick determination as to the amount of ammunition remaining in the box must be made. If there is less than thirty rounds in the box, reload the weapon with a new box or a magazine. The important thing is not to run out of ammunition during the assault. All members of the squad ensure that hand grenades are within easy reach so they can be used during the upcoming assault.

(2) The amount of time the assaulting element spends at the assault position must be kept to the absolute minimum in order to deny the enemy the opportunity to bring fire to bear on the assaulting troops and to keep the momentum of the attack going. It was previously stated that the ideal assault position should offer concealment and cover to the attacking force. What is considered concealment and cover to the attacking force is considered *dead space* by the defender (see par. 2302). Since *dead space* is normally covered by indirect fire from mortars and artillery, the enemy can still bring fire to bear on the assaulting force producing casualties, breaking up the attack, and fixing the attackers in position.
REMEMBER: DO NOT DELAY AT THE ASSAULT POSITION.

b. Final Coordination Line. The final coordination line is used as the control feature to coordinate the ceasing or shifting of the direct and indirect fire which is supporting the assault. Since the assaulting troops will be close to the enemy position, covering fires must be maintained on the enemy as long as possible in order to prevent the enemy from delivering a heavy volume of small arms fire on the assaulting element. The squad must *lean into* the friendly covering fires and get as close to the enemy position as possible before shifting or ceasing these fires. The distance from the final CL to the objective varies with the terrain and types of supporting arms employed. The final CL is tentatively established at the onset of the attack but it is *not a FIXED line*. The assault unit commander will shift or cease fires as the situation and terrain dictate. If the squad leader is the assault element leader and he determines that fewer casualties will be suffered if the squad moves closer to the enemy positions prior to shifting or ceasing covering fires, he may move the final coordination line closer to the enemy positions. The assault element then continues to advance to the new location by rushes or if necessary by creeping or crawling and initiates assault fire from there. When the final CL is changed, the platoon commander and fire support units must be notified. Ideally the assault position and final CL are collocated, but often such may not be the case and the final CL is located between the assault position and the objective. The assault unit commander may have to *walk* the supporting fires in front of his unit before he reaches the final CL where the covering fires are shifted or ceased.

c. Squad in the Assault. The assault must be launched close under the covering fires and begin when the leading assault elements have advanced as close to the enemy as possible without moving into friendly covering fires. The assault is started on order or signal of the platoon commander or on the initiative of the squad or fire team leader. Supporting weapons cover the assault by firing on adjacent or deeper enemy elements. The assault is launched aggressively and vigorously *IMMEDIATELY* upon the shifting from or cessation of covering fires on the objective. The squad advances rapidly and aggressively from the assault position, (deployed in a squad line, fire team skirmishers laying down a heavy volume of assault fire) using assault fire techniques.

(1) Assault Fire. Assault fire is designed to keep the enemy fire suppressed, once covering fires are lifted, by fixing the defenders in their fighting positions. Assault fire permits the assaulting squad to close to within hand grenade range of the enemy position without sustaining heavy casualties from enemy small arms fire. The assault is made as rapidly as possible consistent with the ability of individuals to deliver a heavy volume of well directed fire. The speed of the assault will be governed by the slope and condition of the ground, visibility, and physical condition of the squad. Throughout the assault, fire is directed at every bush or tree stump, every fold in the ground, and every location that might conceivably contain an enemy. Assault fire is characterized by violence, volume, and accuracy. Assault fire is designed to kill and demoralize the enemy, and keep him down until the assault element can overrun the position and kill or capture him.

(a) Riflemen and assistant automatic riflemen advance, delivering well directed shots at locations in their zone of advance. When a definite target appears, it is engaged immediately. Riflemen and assistant automatic riflemen deliver fire by firing from the shoulder using the pointing technique.

(b) Squad automatic riflemen advance while firing their weapons from the underarm position. Fire is delivered in short bursts of three to five rounds. Automatic riflemen distribute their fire across the entire squad front.

(c) In the assault, fire team leaders are fighting leaders. In addition to maintaining control of their teams, they advance while firing the M-203 from the shoulder using the pointing technique. If the M-203 is loaded with high explosive (HE), HE airburst, or HE dual purpose rounds, the weapon should be fired just after reaching the final coordination line in order to suppress enemy fire and to avoid causing friendly casualties. If loaded with a multiprojectile round, firing the M-203 can be delayed until the squad closes with the enemy. Once the M-203 has been fired, the fire team leader fires his rifle until such time as the M-203 can be reloaded. While moving forward in the assault line, it is important to keep up the heavy volume of fire. While reloading the M-203, care must be taken not to fall behind the squad assault line.

(d) The squad leader is also a fighting leader during the assault. He takes a position near the center of the assault line where he can move rapidly to enforce continuity of fire, maintain alignment and momentum, and keep the assault moving forward aggressively.

(2) Decentralization of Control. If the assaulting squad is faced with light enemy opposition, it may be possible for the squad leader to retain control of the unit by maintaining the assault line when sweeping through the objective. However, when enemy opposition is heavy, maintaining the squad skirmish line is not possible. When assaulting a hostile position which is organized in depth, such as a series of emplacements, the squad must attack and destroy, bypass, or contain each enemy position within its assigned zone. When assaulting an organized position, the squad attack often breaks up into a series of separate combats which are continued throughout the depth of the enemy position. Control of the squad under these conditions is very difficult. The importance of quick decision, individual initiative, and speed of action to take advantage of local opportunities requires that control be decentralized during the assault through the hostile position. Under these conditions the squad leader again must rely on the skill and initiative of the fire team leaders and individual Marines to carry out the assigned mission. Each fire team leader and each Marine must take the initiative to use the weapons, grenades, and other ordnance available to the maximum extent possible taking maximum advantage of cover and concealment within the enemy position and employing short, frequent rushes (creeping and crawling when necessary) to close with enemy positions. The first fire team to gain a foothold within the enemy position will support the remainder of the squad in the destruction of the enemy.

(3) Light Enemy Opposition. It was noted that if enemy opposition is light, the squad assault line could be maintained as the squad sweeps through the entire objective. Maintaining the assault line ensures that the squad leader retains control and that the squad will move rapidly through the objective. These advantages must be weighed against the potential danger to the squad. There is a possibility that the enemy will resist until the squad starts the assault and then he will retreat from the objective. As the squad sweeps far enough into the objective to a point where the base of fire is masked by the assault line, the enemy may deliver a heavy volume of fire on the exposed squad, fix it, and then counterattack. The squad leader must consider this danger when determining whether or not to maintain the assault line. It should be remembered that the assault line has served its purpose once the heavy volume of assault fire has pinned the enemy down and the squad has closed with the enemy positions. By directing the fire teams to return to fire and movement at this point, the squad leader can still retain control of the squad, continue a relatively rapid movement through the objective, and reduce the danger to the squad from enemy small arms fire.

4304. Enemy Counterattack

The major concern of the assaulting unit leader once the enemy has been driven from the objective is to retain control of the objective. If the enemy allocated troops to defend the objective in the first place, it is reasonable to assume that he will allocate troops to try to take it back. *It is safe to say that it is not a question of whether or not the enemy will counterattack, but rather a question of when.* In trying to determine when the counterattack will take place, it must be realized that the enemy knows that his chances of success are better if he counterattacks quickly before there is time to build a strong defense. By launching his counterattack quickly, he also knows that the forces now holding the objective will be somewhat disorganized and under strength due to casualties. By striking quickly, the enemy will not give the new defenders time to bring up fresh troops. All things considered, if the enemy acts quickly, his chances of taking the objective back with a relatively small force are better than if he delays while assembling a larger counterattack force. The prudent Marine will expect an enemy counterattack even before the last enemy positions on the objective have been neutralized. Preparations to repel the counterattack must commence immediately after taking the enemy position.

4305. Consolidation

a. General. Consolidation is the rapid organization of a hasty defense in order to permit the attacking unit to hold the objective just seized in the event of an enemy counterattack.

b. Hasty Defense. In receiving the attack order, the squad leader was assigned the mission of seizing and defending an objective or a sector of an objective. The task now is to place sufficient firepower into position to defend that sector. In positioning the fire teams in the hasty defense, there is not sufficient time to prepare standard fighting holes. The squad must use natural depressions, shell craters, or old enemy positions, if available, and quickly improve them to provide minimum adequate cover. This is important since it is expected the enemy will use artillery, mortars, and machine guns to support his counterattack. The emphasis here must be to effectively defend the assigned sector by fire and to get the squad under cover quickly, not perfectly. Fire team sectors of fire are designated and principal directions of fire for automatic rifles are assigned. Each fire team leader must take the initiative to ensure that his team's sector of fire is interlocked with that of adjacent teams. Movement of squad members within the objective should be kept to an absolute minimum in order to reduce exposure to the enemy's artillery, mortar, machine gun, and small arms fire. If a Marine must be moved to a position where he can better cover the fire team sector of fire, he should move by rushes, seeking cover as he moves. During consolidation, there is usually enough time to redistribute ammunition within the fire team. When redistributing ammunition, priority goes to the automatic rifleman. Many of the steps associated with the hasty defense are actually taking place while the enemy is counterattacking. Because of the rapid tempo of events, the full attention of the squad and fire team leaders must be dedicated to the preparation of the hasty defense. Care of casualties must take second priority to the preparation of the hasty defense. Enemy prisoners must be disarmed, searched for other weapons, and guarded. If the squad leader or a fire team leader has become a casualty, the next senior Marine must quickly assume control and carry out the necessary tasks.

(1) Light Enemy Opposition. When the enemy resistance on the objective is light, the squad will remain relatively intact and under the squad leader's control as it moves through the objective. Under these conditions, squad and fire team leaders can commence the development of the hasty defense in an orderly fashion. Light enemy opposition on the objective does not reduce the danger of enemy counterattack. It may be an indication, since his forces have not suffered heavy casualties in the fight for the objective, that he will launch a stronger counterattack. The need to rapidly develop the hasty defense still exists.

(2) Heavy Enemy Opposition. Paragraph 4303c(2) described the difficulty a squad leader encounters in maintaining control of the squad when enemy resistance on the objective is heavy. When the squad assault breaks up into a series of individual combats the squad leader's task of building a hasty defense is made even more difficult because the squad's organizational integrity has temporarily broken down. The squad and fire team leaders are now faced with the task of building a hasty defense without having the usual control of the unit. In this situation the squad leader proceeds though the objective, gathering as many members of his squad as possible and positioning them to cover the assigned defensive sector. He ensures that his men take maximum advantage of available cover. If he has only one of his fire teams intact, it may be necessary to assign this team to cover the entire squad sector. Usually, due to the confusion of the battle for the objective, there will be individual Marines in the squad sector who have been separated from their fire teams. These Marines may be from another squad or even another platoon. If they appear to be *strays* and are in the squad sector, the squad leader takes charge of these Marines, assigns them a sector of fire, and gets them into a covered position. The squad leader may temporarily assign *strays* to fire teams already in position or if necessary, form temporary fire teams. If a Marine is wandering around looking for his fire team, his fire power will be lost to the unit when the enemy counterattacks; if he remains exposed, he will most likely become a casualty. *During consolidation, the primary task is building the hasty defense, not reforming the squad.* Unit leaders must make their presence felt when building the hasty defense to assure each Marine that he is not alone in resisting the enemy counterattack. When enemy resistance on the objective is heavy, the fire team leader should remain close to the automatic rifleman and attempt to get automatic riflemen through the objective and integrated into the hasty defense as quickly as possible. The fire power of the automatic rifleman is critical to the hasty defense.

(3) Termination. Hasty defensive positions are maintained until the reorganization of the unit commences.

4306. Reorganization

Once the enemy counterattack has been defeated or it has been determined by the senior unit leader on the objective that the danger from immediate enemy counterattack has passed, reorganization of units commences. Reorganization is a continuous process, but it is given special emphasis upon seizure of the objective. The squad leader accomplishes the following during reorganization:

- Makes spot assignments to replace fire team leaders and automatic riflemen who have become casualties.

- Redistributes ammunition, magazines, and grenades.

- Removes casualties to covered positions.

- Notifies the platoon commander of the situation, the position of the squad, the casualties incurred, and the status of ammunition supply.

- Delivers enemy prisoners to the platoon commander. Prisoners and enemy dead are searched for weapons, papers, documents, and identification. Such material is immediately sent to the platoon commander. (See app. G.)

- Ascertains the situation of the units to his flanks.

Section IV. Exploitation Phase

4401. Exploitation

Exploitation normally occurs after a successful assault and seizure of the objective. It begins immediately after or in conjunction with the consolidation and reorganization phase. It is a continuation of the attack aimed at destroying the enemy's ability to conduct an orderly withdrawal or organize a defense. Pursuit by fire and/or continuation of the attack are methods used to exploit success.

a. Pursuit by Fire. When the assault through the assigned objective is completed, the squad fires upon the withdrawing enemy forces until they are no longer visible or are beyond effective range.

b. Continuation of the Attack. The purpose of continuing the attack is to maintain pressure on the retreating enemy and destroy his combat power. When ordered, the rifle squad continues the attack. The squad leader repeats all the steps performed for previous attacks. Frequently, the urgent need of a higher command to maintain momentum requires that these steps be done rapidly so that the attack can be continued with minimum delay.

Section V. Night Attack

4501. General

a. Purpose. A night attack may be made to gain surprise, to maintain pressure, to exploit a success in continuation of daylight operations, to seize terrain for subsequent operations, or to avoid heavy losses by using the concealment afforded by darkness.

b. Characteristics. Night combat is characterized by a decrease in the ability to place aimed fire on the enemy; a corresponding increase in the importance of close combat, volume of fire, and the fires of weapons registered during daylight; difficulty of movement; and difficulty in maintaining control, direction, and contact. Despite these difficulties, the night attack gives the attacker a psychological advantage in that it magnifies the defender's doubts, apprehensions, and fear of the unknown. The difficulties mentioned can be overcome by careful planning and preparation for the attack. The demand for time-consuming detailed planning and reconnaissance at all levels normally requires the assignment of night attack missions to units not in physical contact with the enemy.

4502. Tactical Control Measures

The degree of visibility will determine the measures necessary to assure control. Terrain features used as tactical control measures, if not easily identifiable at night, may be marked by artificial means. The following control measures will normally be prescribed in a night attack. (See fig. 4-10.)

a. Assembly Area. The assembly area may be closer to the line of departure than for a daylight attack.

b. Attack Position. The attack position should be in defilade, but need not offer as much concealment as in daylight. The area selected should be easy to move into and out of at night.

c. Line of Departure. The line of departure is a line established to coordinate attacking units when beginning the attack.

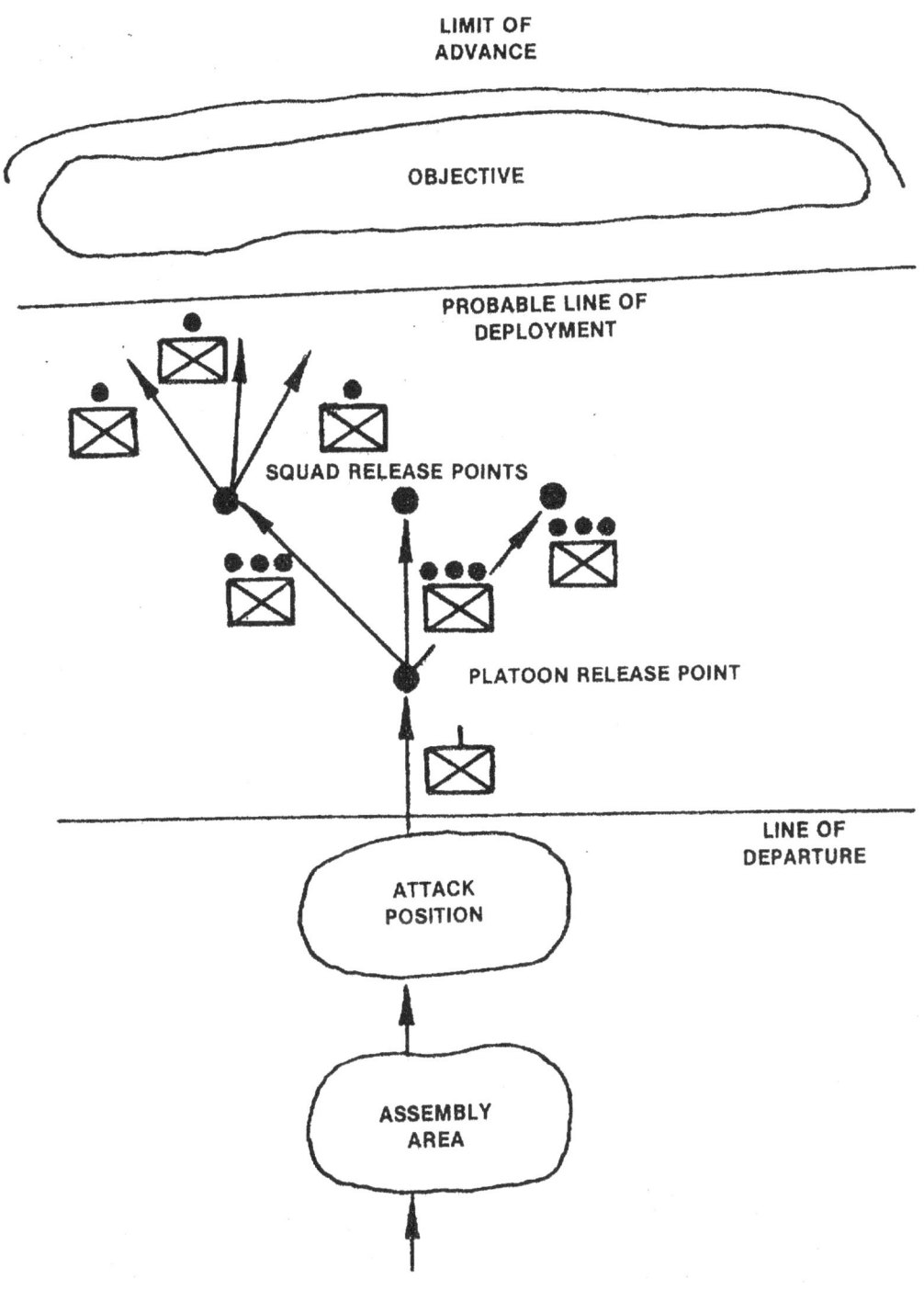

Figure 4-10. Control Measures for Night Attacks.

d. Release Point. Release points are clearly defined points on a route where units are released to the control of their respective leaders.

e. Probable Line of Deployment (PLD). The probable line of deployment is an easily recognized line selected on the ground where attacking units deploy in line formation prior to beginning a night attack.

f. Limit of Advance. A line of advance is generally designated beyond the objective to stop the advance of attacking units. It should be easily recognizable in the dark (a stream, road, edge of woods, etc.) and far enough beyond the objective to allow security elements space to operate.

4503. Security Patrols

Members of the squad may be used as security patrols to assist night attacks. These patrols maintain security on the PLD, eliminate enemy security elements, and prevent attacking forces from being ambushed while en route to the PLD. They may also act as guides to lead units forward from the release points to the PLD.

4504. Preparatory Phase for the Night Attack

Preparation for the night attack is generally the same as for the daylight attack. The squad leader will follow the same format outlined in appendix C when preparing for and executing assigned missions. Special emphasis is placed on —

- Reconnaissance by squad and fire team leaders to locate assigned control features and terrain features for night orientation. Such reconnaissance should be conducted during three conditions of visibility: daylight, dusk, and dark.

- Rehearsals conducted both during daylight and darkness. Rehearsals should include formations, audible and visual signals, and the actions of the squad from the assembly area to the objective.

- Carrying only that equipment absolutely essential for the success of the attack.

- Camouflaging individuals and equipment. Equipment which rattles is padded or tied down.

- Avoiding test firing of weapons and unnecessary movement, or doing this in a way which will not prematurely disclose the forthcoming attack.

- Ensuring that the night vision of the squad members is not destroyed prior to the attack.

4505. Conduct Phase of the Night Attack

a. Movement to the Probable Line of Deployment

(1) Security patrols sent out by higher commanders destroy enemy listening posts and security patrols enabling the unit to move to the PLD undetected.

(2) The platoons move in column formation from the assembly area to the platoon release point. At the platoon release point, the platoons meet their guides from the security patrol and continue to move along their respective routes to their squad release points.

(3) Once the squad crosses the line of departure, movement to the PLD is continuous. The rate of advance is slow enough to permit silent movement.

(4) If flares are fired during the movement forward, all hands quickly assume the prone position until the flares burn out. If a flare is fired after the squad leaves the PLD, the squad ignores the flare and continues the movement toward the assigned objective. Close coordination is required on the use of flares. Indiscriminate use of flares results in loss of surprise. If the attack is to be illuminated, the illumination is started on signal from the attacking elements (usually after reaching the PLD).

(5) On arrival at the squad release point, the rifle squads are released from the platoon column formation to deploy on line at the probable line of deployment. The squad leader is normally the first member of his squad in the column. When the rifle squad reaches the squad release point, he leads the column, sets the pace, and maintains the direction of movement. Members of the security patrols assist the squad leaders in positioning the squads on the probable line of deployment.

(6) On order, the squad moves forward silently from the PLD, maintaining the squad line formation and guiding on the base squad.

b. Assault. Once the enemy has discovered the attack and firing has commenced, then and only then is the assault commenced. The signal for the assault can take any form, but it must be simple and reliable. The importance of developing a great volume of fire during the assault cannot be overemphasized. It is at this time that fire superiority must be established and maintained. The assault is conducted aggressively. Tracer fires should be used to increase accuracy of fire and to demoralize the enemy. Preplanned fires are used by higher commanders to isolate the objective. The assault is conducted in the same manner as discussed in paragraph 4303. The assault is carried forward to the forward military crest of the objective or to some other prescribed limit, short of the limit of advance.

4506. Consolidation and Reorganization Phase of the Night Attack

When the objective has been seized, the plans for consolidation and reorganization are carried out as described in paragraphs 4305 and 4306. The squad does not move or employ security elements forward of the limit of advance (see fig. 4-10) until ordered.

Section VI. Infiltration

4601. General

Infiltration is a technique by which a force moves as individuals or small groups over, through, or around enemy positions without detection. Although primarily offensive in nature, an infiltration can be conducted in conjunction with defensive or retrograde operations. The purpose of an infiltration is to gain a more favorable tactical position from which to perform a subsequent mission.

4602. Planning and Preparation

a. Organization. The size of the infiltrating group depends primarily on the need for control between infiltrating groups, and the number and size of the gaps in enemy defenses. Normally, units will be broken down into infiltration groups of platoon or squad size.

b. Order. A detailed order is issued for the infiltration. Each infiltrating group is issued the following information as a minimum:

- Release point.
- Time of release.
- Point of departure.
- Time of infiltration.
- Infiltration lane.
- Rendezvous point.
- Alternate rendezvous points.
- Time of rendezvous.
- Routes from rendezvous to attack positions.

c. Preparation. Upon receipt of the order, the infiltration group leaders follow the troop leading procedures as discussed in appendix C. While the group leaders accomplish their troop leading steps, their assistants prepare their groups for infiltration. Necessary equipment is drawn, checked, and secured for silent movement. Each man prepares himself and his equipment for the operation. Whenever possible, each infiltration group should carry the necessary special equipment to accomplish the mission of the infiltration force. This ensures the accomplishment of the mission in the event all groups do not successfully complete the infiltration. After the group leaders issue their orders, rehearsals are conducted. Rehearsals should address the passage of lines, signals, actions at danger areas, actions upon enemy contact, and actions to be taken at the rendezvous points and the objective. Everyone should be required to memorize the route, azimuths to, and location of rendezvous points. The accomplishment of the mission rests primarily on the ability of the small unit leaders. The planning and preparation must be as thorough and as detailed as time and facilities will permit. Fires are planned by higher headquarters to create diversions and to protect and support the unit during the infiltration, in the rendezvous area, and during any subsequent attack, consolidation, or withdrawal.

d. Control Measures

(1) Infiltration Lanes. Infiltration lanes extend through known or likely gaps in the enemy defenses and are often located in rough, swampy, or heavily forested areas. (See fig. 4-11.)

(2) Rendezvous Points. Rendezvous points should be concealed from possible detection by enemy observation and patrols. They are secured by the first group into the area. Escape routes should be designated to alternate rendezvous points.

(3) Time of Infiltration. The time of infiltration is selected to take advantage of conditions of reduced visibility, such as darkness, rain, snow, fog, and so forth. It is the time when infiltration groups enter their assigned infiltration lanes.

(4) Routes. Routes to the objective from the rendezvous points should be concealed for surprise and for protection.

(5) Objectives. Objectives may be enemy reserves, artillery units, or command and logistic installations. Infiltrating forces may also seize key terrain or establish roadblocks to restrict enemy movement, isolate the battle area, and facilitate the movement of friendly mechanized forces.

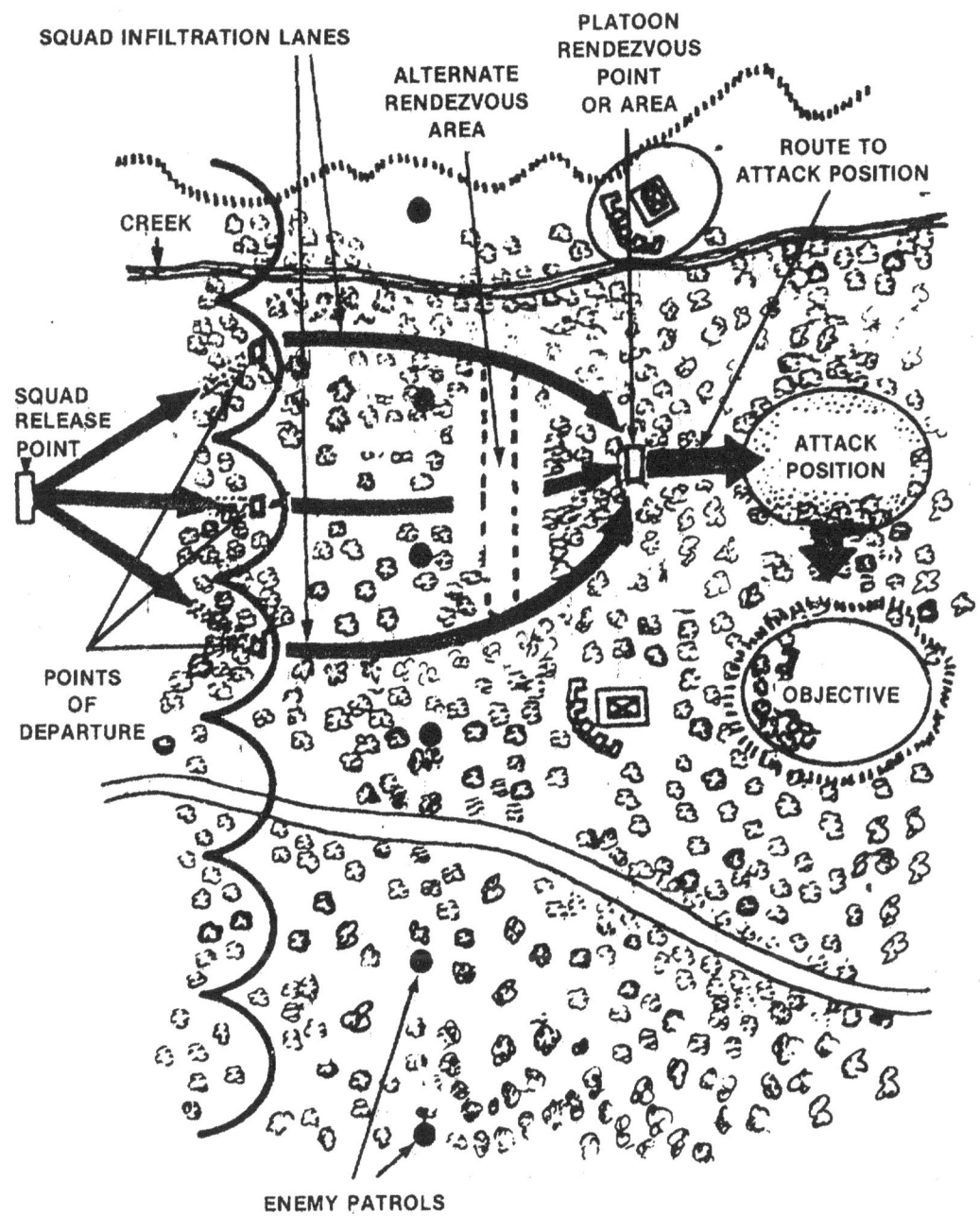

Figure 4-11. Attack Using Infiltration Techniques.

4603. Conduct of the Attack by Infiltration

a. Movement of Groups. The unit conducting the infiltration will assemble the infiltration groups to the rear of friendly lines. The unit will then move forward, usually in a column, until it reaches the release point. At the release point, the infiltration groups are released to their leaders. The infiltration groups move by stealth to avoid detection. They cross the line of departure (usually friendly frontlines) at the specified time, normally during darkness. Artillery or mortar fires are used as necessary to distract enemy attention. The infiltration groups pass through the gaps in the enemy lines by using the infiltration lanes. If detected, groups avoid engagement by withdrawing or moving around the enemy. Speed of movement is limited by the requirement for stealth. Groups which are unable to reach their rendezvous point in time follow the previously announced alternate plan.

b. Assembly of Groups. At the rendezvous point, groups assemble and assault preparations are completed. The first infiltration group to reach the rendezvous point secures it. The assembled force leaves the rendezvous point to assault the objective at the designated time. The main body may be preceded by a small security element (scouting fire team). Its mission is to prevent the main body from being detected or surprised.

c. Assault. The force is halted short of the objective for final reconnaissance and coordination. This assault position should be the last safe, covered, and concealed area before reaching the objective. The assault on the objective is characterized by surprise and maximum firepower at the objective's weakest point to quickly destroy or capture it. If plans are to link up with other friendly forces, previously designated visual and sound recognition signals prevent fire fights between friendly units. If the objective is not retained, the assaulting force withdraws to an assembly area for further assaults or withdraws to friendly lines. The withdrawal to friendly lines may be by air or by exfiltration, either as an intact unit or by exfiltration groups. Upon reaching friendly lines, the unit is again reassembled.

Chapter 5

Defensive Combat

Section I. General

5101. Purpose

The purpose of defensive action is to retain or control terrain, gain time, develop more favorable conditions for offensive action, or to economize forces to allow the concentration of forces elsewhere.

5102. Mission

The mission of the infantry in the defense is, with the support of other arms, to stop the enemy by fire as he approaches the battle position, to repel his assault by close combat if he reaches the battle position, and to destroy him by counterattack if he enters the battle position. For the rifle squad, this mission can be divided into three parts:

a. To destroy the enemy by fire once he comes into small arms range of the squad's fighting position. The enemy is destroyed as far forward of the squad's fighting position as possible. The closer the enemy comes to the squad's fighting position, the more friendly casualties he will inflict.

b. If the enemy continues to press the attack to the point where he launches an assault, the squad repels this assault by continuing to deliver fire as part of their unit's final protective fires and, if necessary, by hand-to-hand combat.

c. If the enemy succeeds in penetrating the platoon battle position, the squad holds its fighting position, delivering fire on the intruding enemy and participating in counterattacks to destroy the enemy and restore the battle position.

5103. Definitions

a. Sector of Fire. A sector of fire is an area which is required to be covered by fire by an individual, a fire unit (squad or fire team), or a crew-served weapon. It is a pie-shaped area enclosed by two lateral limits and a forward limit. (See fig. 5-1.) Within a rifle platoon, a sector of fire is assigned to individual weapons, fire teams, and squads. Squad leaders are not normally assigned individual sectors of fire since their primary duty during the conduct of the defense is directing and controlling the fires of their units. The sector of fire is used to clearly indicate the area to be covered by fire and to provide for the best distribution of available firepower and complete coverage of the area to the front. It is also employed to ensure mutual support by the overlapping of adjacent sectors of fire. Rifle platoons are assigned battle positions to be defended. The rifle platoon battle position is defended by the overlapping sectors of fire of the squads. The squad sector of fire is covered by the overlapping sectors of fire of the fire teams.

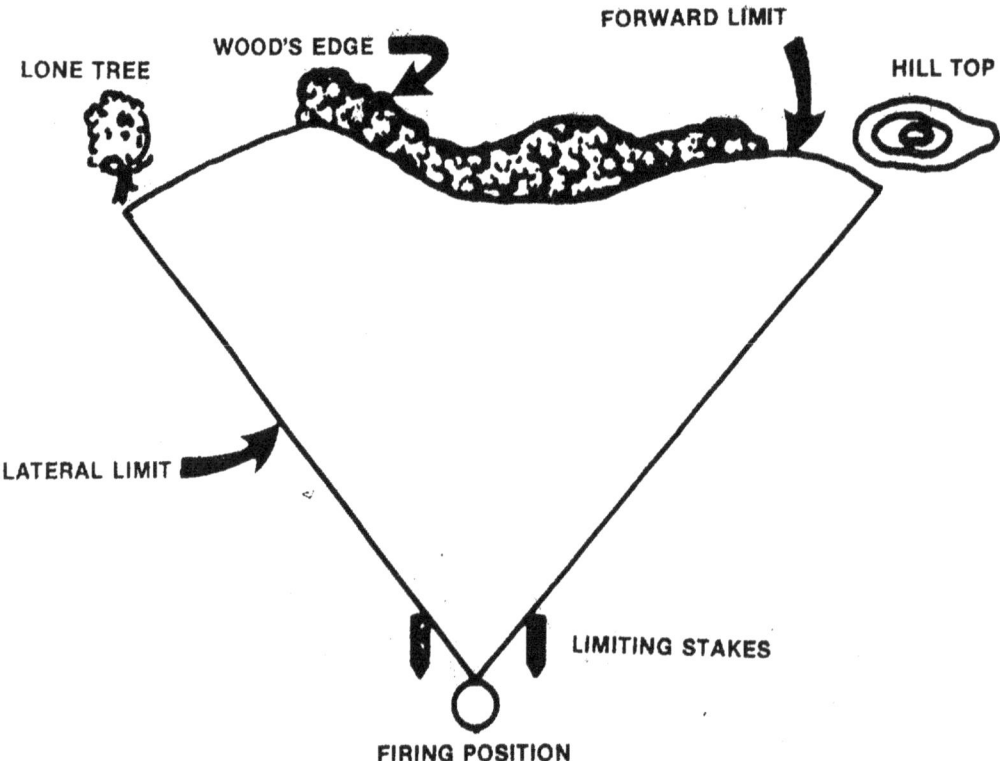

Figure 5-1. Sector of Fire.

(1) Lateral Limits. Readily identifiable terrain features are selected to indicate the line of sight along each side of the sector. These terrain features should be located near the forward limit of the sector so that all fire team members assigned to cover this sector use the same limiting features. Two stakes, placed near the position of the weapon, are used to indicate the lateral limits during periods of restricted visibility. These should be emplaced prior to darkness.

(2) Forward Limit. The forward limit is established at the range at which the weapon will open fire. For rifles and automatic rifles, this may extend up to their maximum effective ranges. When possible, a terrain feature is selected to locate the forward limit. As the attacker passes this limit, he is brought under fire. This allows the squad leader a positive means to control the commencement of small arms fire.

b. Fighting Position. A fighting position is a location on the ground from which fire is delivered by an individual, a fire unit (squad or fire team), or a crew-served weapon. Before selecting a firing position, the assigned sector of fire must be carefully examined from various locations using the prone position to ensure effective coverage of the sector of fire. The exact fighting position is then designated on the ground prior to digging in. The position must allow for good fields of fire, make maximum use of available cover and concealment, and facilitate exercise of fire control by the unit leader.

(1) Primary Fighting Position. The primary position is the best available position from which the assigned sector of fire can be covered. Individuals, fire teams, squads, and crew-served weapons are assigned primary positions.

(2) Alternate Fighting Position. Alternate positions are not normally assigned to individuals or units within the platoon. They are used primarily by crew-served weapons. An alternate position is located so that a crew-served weapon can continue to accomplish its original mission when the primary position becomes untenable or unsuited for carrying out that mission.

(3) Supplementary Fighting Position. One of the greatest threats to either the attacker or the defender lies in being surprised. The attacker seeks to surprise the defender by concealing his movements until the moment of the assault. The defender seeks to surprise the attacker by concealing the exact location and extent of his dispositions, thus leading his opponent into a false estimate of the situation and consequently, a faulty decision. Supplementary positions are prepared to guard against attack from directions other than those from which the main attack is expected. A supplementary position is a secondary position and does not cover the same sector of fire as the primary position. *In some situations, the most likely avenue of approach may vary between daylight and darkness or other periods of low visibility. Thus, the requirement to shift positions becomes an absolute necessity. This situation is more the rule rather than the exception.* Supplementary positions actually provide security. When occupied, they ensure protection against attack from directions other than those covered by primary positions. Movement to supplementary positions should be made by covered and concealed routes when available.

c. Battle Position. A position on which the main effort of defense is concentrated. A battle position is assigned to battalions, companies, and platoons. A battle position is made up of a series of sectors of fire that support one another. Platoon battle positions are assigned a right and a left limit of fire. A limit of fire is a boundary marking the area in which gunfire can be delivered. The limits of fire should be indicated by readily identifiable terrain features located at or beyond the limit of effective small arms fire. (See fig. 5-2.)

d. Forward Edge of the Battle Area (FEBA). The foremost limits of a series of areas in which ground combat units are deployed. The FEBA is a control measure that divides the security area from the main battle area. The FEBA need not be physically occupied, but it should be controlled by friendly fire.

e. Main Battle Area. The main battle area extends from the FEBA to the rear. It is here that the decisive defensive battle is fought. The main battle area is organized into sectors of defense that are assigned to subordinate units. A company may be assigned a sector to defend or be directed to occupy a battle position.

f. Security Area. The security area is located forward of the FEBA. The squad may be assigned as part of a larger security force or may only be responsible for local security in front of the platoon battle position.

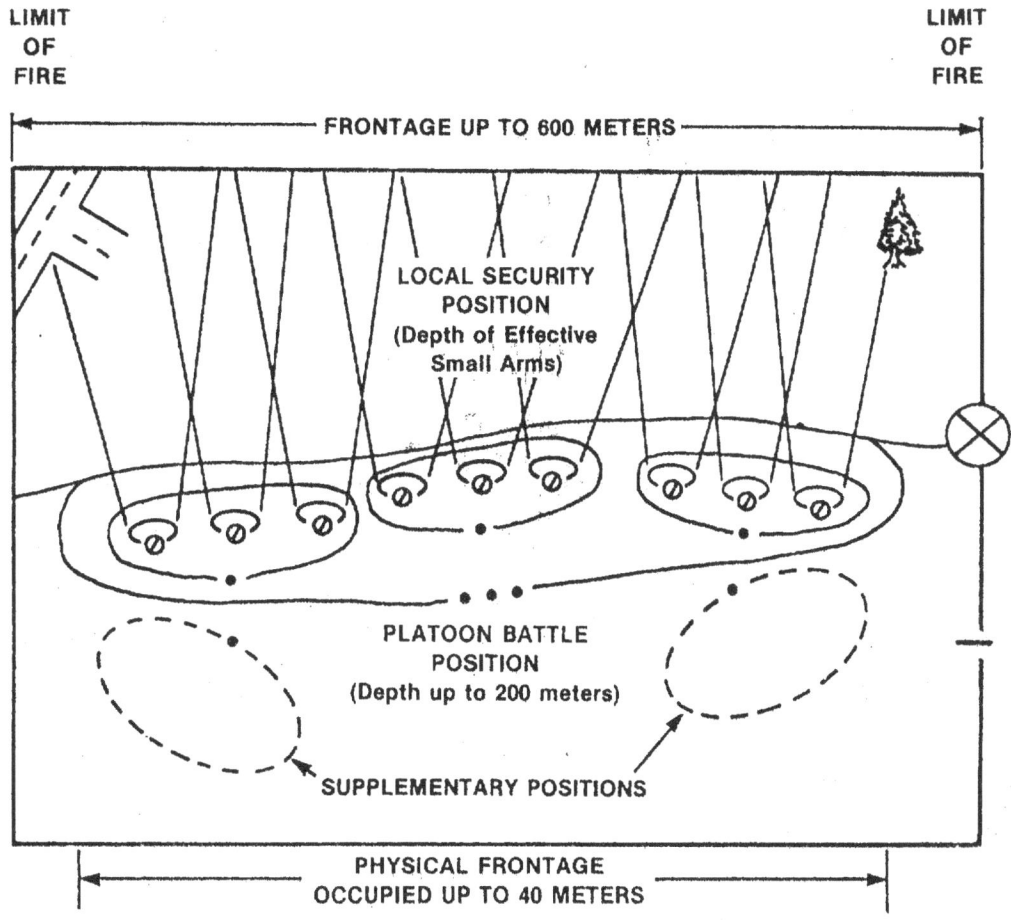

Figure 5-2. Frontline Platoon Battle Position.

g. Principal Direction of Fire. A principal direction of fire is a specific direction within the sector of fire given to a flat-trajectory weapon and which is designated as its primary fire mission. Within a rifle squad, a principal direction of fire is assigned to automatic rifles. Units are not assigned principal directions of fire. Riflemen may be assigned principal directions of fire for periods of reduced visibility. Squad leaders and fire team leaders are not assigned a principal direction of fire, nor can an automatic rifle be assigned more than one principal direction of fire. The principal direction of fire is indicated by pointing out a readily identifiable terrain feature.

This terrain feature may be the target itself or it may indicate the line of sight when no target is assigned. The limits of the target should be pointed out on the ground when distributed fire is required along the principal direction of fire. A stake near the firing position is used to indicate the principal direction of fire during periods of restricted visibility. The principal direction of fire is employed to —

- Cover a gap in a final protective line of a machine gun.

- Cover a specific terrain feature endangering the company or platoon battle position, such as a draw which may serve as an avenue of approach, or hill top which may serve as a possible enemy vantage point. (See fig. 5-3.) This terrain feature is not necessarily a point on which fixed fire is placed; however, it is intended that coverage of the feature should require little distribution of fire. The principal direction of fire may be anywhere within the sector of fire.

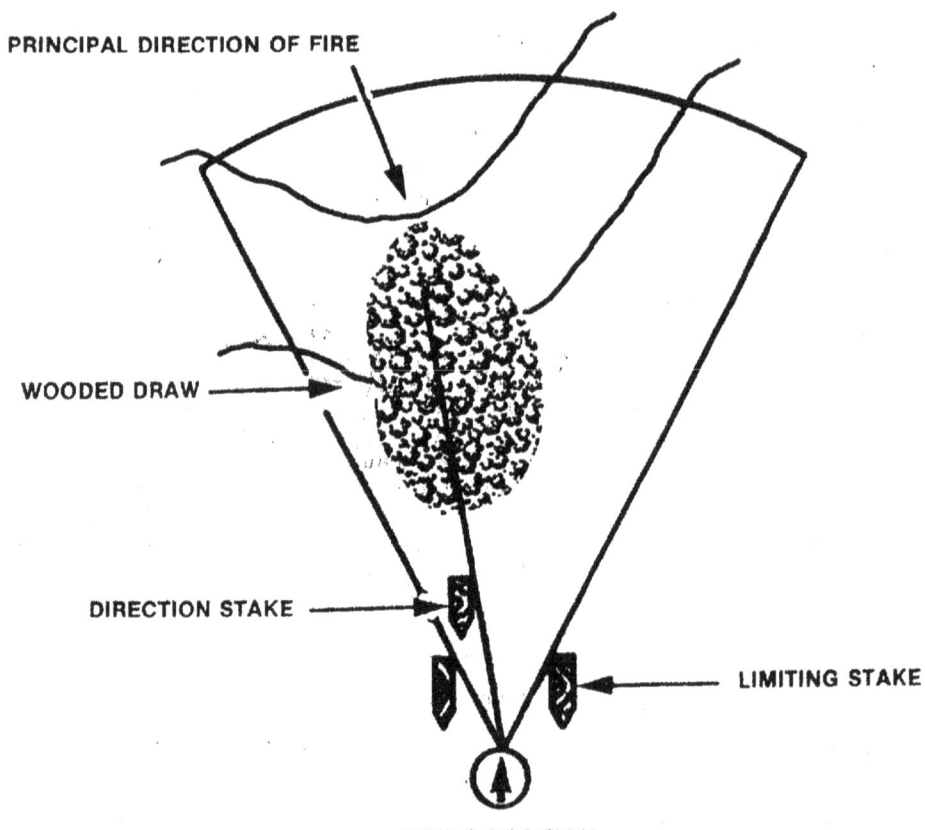

Figure 5-3. Principal Direction of Fire.

- Protect a crew-served weapon by firing across its front.

- Augment the band of flanking fires placed immediately in front of the battle position when targets of opportunity to the front are not visible.

5104. Fundamentals of Defense

The following fundamentals of defense are applicable to all tactical levels (fire team, squad, platoon, company, etc.). Application of these fundamentals by all unit leaders increases the chances for a successful defense.

a. Preparation. Normally, the defender will arrive at the battlefield before the attacker. Upon his arrival at the position he is to defend, the squad leader must ascertain from the platoon commander how much time is available to prepare his defensive position. If time is available to prepare the position the squad leader must use it wisely; if not, he prepares a hasty defense.

b. Concentration. Forces must be concentrated to prepare for attacks at the most likely spots. For the squad leader, this means he will establish his position as directed by the platoon commander.

c. Flexibility. At the squad level, flexibility is achieved through the continuous development of various courses of action to meet the enemy threat. The squad leader should continuously be asking himself, "What do I do if the enemy does this. . . ?"

d. Maximum Use of Offensive Action. The squad will normally be tasked by the platoon commander to conduct various types of patrols to maintain contact with the enemy. Additionally, the squad leader must instill in his men an offensive state of mind and aggressive spirit. Collecting extra equipment to provide for *creature comforts* must not be allowed as it can affect the physical and mental ability of the squad to move out quickly and aggressively.

e. Proper Use of Terrain. Take maximum advantage of the military aspects of terrain – key terrain, observation, cover and concealment, obstacles, and avenues of approach (KOCOA).

f. Mutual Support. Units and supporting weapons are located and employed so that they can assist one another. An isolated unit is easily destroyed by the enemy. Positions should be located so that when attacking one, the enemy comes under fire from at least one other.

g. Defense in Depth. The squad employs all three fire teams on line when deployed. The squad engages the enemy at maximum small arms range and continues to fire until the enemy is stopped. If the attacker penetrates the frontline squads, those squads may move to supplementary positions to continue to engage the enemy, or they may be part of a counterattack to drive the enemy back.

h. Surprise. The squad leader must employ every means available to mislead the enemy as to the true locations of his positions, his strength, and the disposition of his organic weapons as well as any crew-served weapons located in his sector.

i. Knowledge of the Enemy. Since the defense reacts largely to what the attacker does, the squad leader should find out from the platoon commander, the capabilities of the enemy facing him. Having an idea as to what the enemy can do, what weapons he will employ, and what his strength is, will help the squad leader organize his defense to meet that threat.

5105. Defensive Missions of the Squad

The squad will be assigned one of three types of missions:

a. Frontline Squad. The squad may defend as part of a frontline platoon. Its mission is to stop the enemy by fire forward of the platoon battle position and to repel him by close combat if he reaches the platoon battle position. The mission requires that the squad be assigned a fighting position and a sector of fire. The squad holds its fighting position at all costs and withdraws or occupies other fighting positions only on orders from higher authority.

b. Squad as Part of the Reserves. The squad may be part of the reserve platoon during the defense. As part of the reserve platoon, the squad is normally assigned a fighting position to the rear of the frontline units and supports them by fire. The fighting position and sector of fire is assigned to concentrate fire in the rear, on the flanks, or into a gap between frontline platoons. The squad as part of the reserve platoon may also be assigned a fighting position and sector of fire to limit enemy penetrations of the platoon battle position. The squad as part of the company's reserve platoon may participate in a counterattack to expel the enemy from the company battle position.

c. Squad as a Security Element. During the defense the squad may serve as part of the security element located forward of the platoon battle position. The squad's mission in this capacity is to gain information about the enemy and to deceive, delay, and disorganize his advance.

5106. The Fire Team in the Defense

a. Organization of the Ground. The squad leader organizes the fire team in the defense by specifying a sector of fire and principal direction of fire for the automatic rifle. He selects terrain features to indicate lateral and forward limits of the sector of fire. He points out on the ground the general location of fire team fighting positions to be occupied. (See fig. 5-4.)

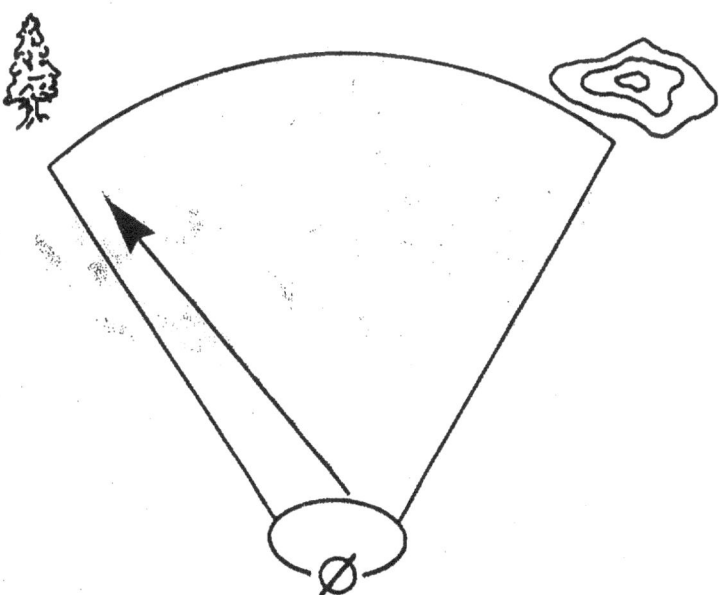

Figure 5-4. The Fire Team Sector of Fire.

b. Fire Plan. The fire team leader formulates the team's fire plan to cover the entire sector assigned by the squad leader with the heaviest possible volume of fire. (See fig. 5-5.) The fire plan includes assignment of individual sectors of fire, individual fighting positions, firing positions and a principal direction of fire for the automatic rifle as assigned by the squad leader, and the position of the fire team leader.

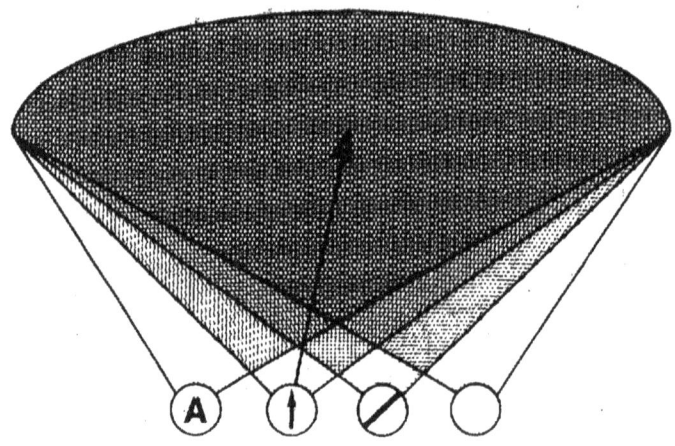

Figure 5-5. Fire Team Fire Plan.

(1) Individual Sectors of Fire

(a) The fire team is the basic fire unit of the rifle platoon and, when practicable, each individual's sector of fire covers the entire fire team sector of fire. The same terrain features are used to indicate the limits.

(b) In the defense, it is impractical for each automatic rifleman to cover the entire squad sector of fire. He is assigned to cover only the fire team sector.

(c) The fire team leader is assigned an individual sector of fire for the employment of the M-203 grenade launcher. He covers the entire fire team sector. The fire team leader normally does not fire his rifle except in emergencies since his primary duty during the conduct of the defense is to control and direct the fire of his team, particularly the automatic rifle.

(2) Individual Fighting Position

(a) The fire team leader designates individual fighting positions which will enable the fire team to cover the assigned sector by fire.

(b) Positions may be prepared as single or double fighting holes. The interval between fighting holes within a fire team may vary from 5 to 20 meters. In close terrain, single fighting holes are usually prepared.

(c) If double fighting holes are prepared, the automatic rifleman and assistant automatic rifleman will pair off.

(3) Automatic Rifleman

(a) Since the automatic rifles are the backbone of the squad's defense, the squad leader selects the exact fighting position for the automatic rifle. The remainder of the fire team is then positioned around it.

(b) The squad leader will indicate the principal direction of fire for the automatic rifle. This principal direction of fire, under some conditions, may have been selected by the platoon commander himself. (See par. 5107a.)

(4) **Rifleman.** The rifleman is positioned so he can cover the entire fire team sector, if possible. His position must provide support and protection for the automatic rifleman.

(5) **Assistant Automatic Rifleman.** Normally, the assistant automatic rifleman participates in the defense as a rifleman. He is positioned near or with the automatic rifleman because he must be prepared to assume the duties of the automatic rifleman.

(6) **Position of the Fire Team Leader.** Usually the fire team leader's position is at the center of the fire team. It must be a position from which he can—

- Observe the entire fire team and its sector of fire.
- Direct the fire of the automatic rifle.
- Deliver effective M-203 grenade launcher fire.
- Observe the squad leader, if possible.

(7) M-203 Employment

(a) In assigning the sectors of fire for employing his M-203 grenade launcher, the fire team leader must consider the overall fire plan. Specifically, he must consider the sectors of fire assigned to the automatic rifleman and the need to furnish support to the automatic rifleman and to adjacent units. The fire team leader then positions himself where he can best control the fire team and deliver the most effective M-203 fire. This is usually in the center of the fire team position. (See fig. 5-5.)

(b) As the enemy approaches the platoon battle position, he is subjected to an ever-increasing volume of fire from weapons in the battle position and from supporting arms. Unless restrictions are placed on the firing of the M-203, the fire team leader opens fire with the M-203 on profitable targets as they come in range. In some situations, the squad leader or platoon commander may desire to withhold M-203 fires until the enemy has reached a specified area, at which time the fire team leader opens fire. The surprise fire from the grenade launcher, in conjunction with the fires of the other squad and platoon weapons, will have a devastating effect upon the enemy, particularly in the assault phase of the enemy attack. When final protective fires are called for, the fire team leader engages the largest mass of enemy infantry within his assigned sector with the M-203.

(c) The fire team leader's fighting position should enable him to cover the entire fire team sector of fire. Primary and supplementary fighting positions are prepared. Firing positions are selected to provide maximum cover and concealment consistent with the assigned mission. Extreme care must be taken to ensure that fields of fire are cleared of obstructions that might cause premature detonation of the projectile, thereby endangering friendly personnel. The M-203 is employed to cover the most likely avenues of approach for enemy infantry into the defensive position.

(8) Fire Team Sectors of Fire. Sectors of fire are selected for the fire teams so that when combined they will cover the entire squad sector of fire. The fire team sectors of fire overlap so as to provide mutual support.

(9) Fire Plan Sketch. A sketch of the fire plan is submitted by the fire team leader to the squad leader. It should include the individual sectors of fire and primary fighting positions, the principal direction of fire for the automatic rifleman, and the fighting position of the fire team leader. At times, irregularities within the terrain may prevent one of the individuals from covering the entire fire team sector of fire. Such is the case of the assistant automatic rifleman in the example shown in figure 5-6. Note that the symbol for the automatic rifleman's fighting position points along the principal direction of fire. The magnetic north line provides a reference to show the direction the fire team is facing. A line is drawn around the fire team fighting position and follows the general trace of the forward edge, flanks, and rear of the individual fighting positions of the fire team members. The symbol indicating the size of the unit is placed within a break along the rear edge. The numbers show this to be the 2d fire team of the 3d squad. Figure 5-7 illustrates the meaning of the various symbols.

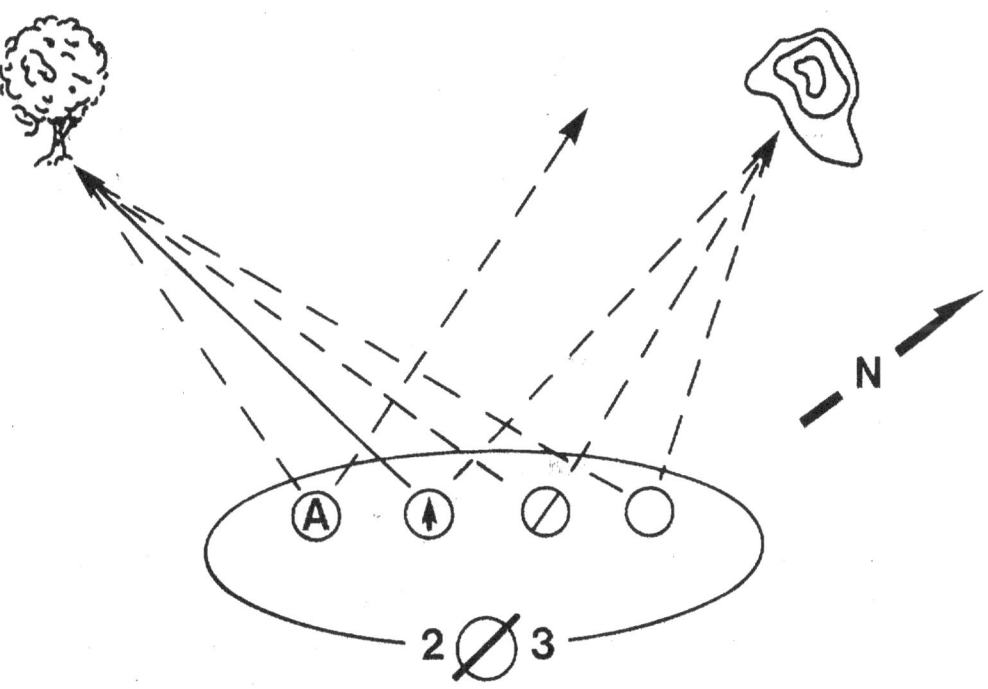

Figure 5-6. Fire Plan Sketch.

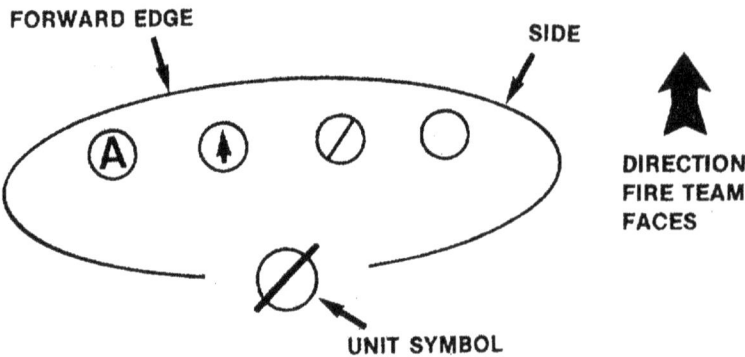

Figure 5-7. Sketch Symbols.

5107. The Rifle Squad in the Defense

a. Organization of the Ground. The platoon commander organizes the fire team in the defense by specifying a sector of fire and a primary fighting position. He selects terrain features to indicate the lateral and forward limits of the squad sector of fire. He points out, on the ground, the general location of the squad fighting position to be occupied. He designates the general fighting positions and principal directions of fire for specific automatic rifles which are critical to the defense of the entire platoon. He may assign a supplementary fighting position for the squad to protect the flanks or rear of the platoon battle position. (See fig. 5-8.)

b. Fire Plan. The squad leader formulates the squad fire plan so as to physically occupy the assigned primary fighting position and to be able to cover by fire the sector of fire assigned by the platoon commander. (See fig. 5-9.) The fire plan includes the assignment of fire team sectors of fire, fire team fighting positions, principal directions of fire for the automatic rifles, and the squad leader's fighting position.

c. Fire Team Positions

(1) The squad leader distributes his fire teams so that they physically occupy the assigned fighting position and are able to cover by fire the assigned squad sector of fire.

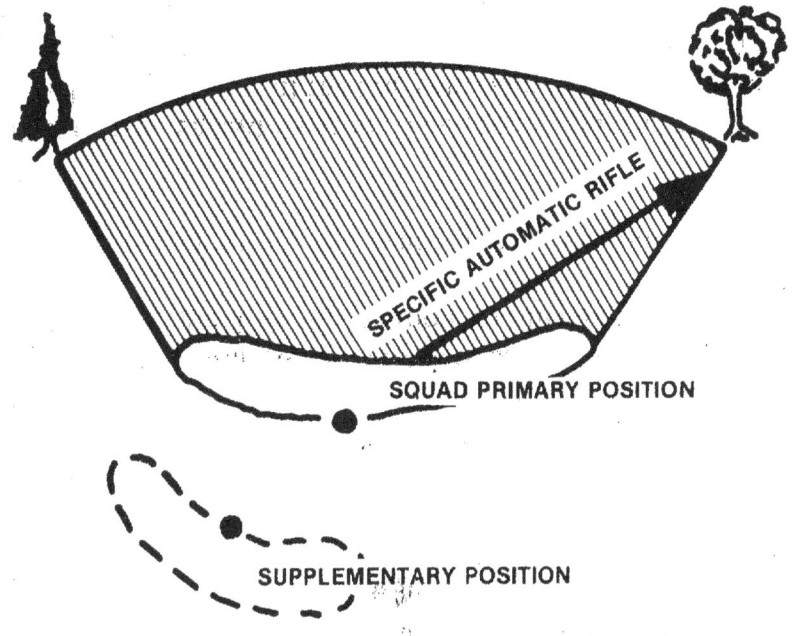

(ONLY ONE OF THE THREE SQUAD AUTOMATIC RIFLES ARE CRITICAL TO THE DEFENSE OF THE ENTIRE PLATOON.)

Figure 5-8. Squad Fighting Position.

(2) Fire teams are placed generally abreast. They face the expected direction of enemy attack so as to be able to deliver their heaviest volume of fire against the enemy forward of the platoon battle position. Fighting positions of individual fire team members may be staggered in an irregular line to take advantage of the terrain; however, care must be taken not to mask the fires of members of the fire team.

(3) Selection of fire team fighting positions must be coordinated with the location of crew-served weapons in the squad fighting position so as to provide for the close-in protection of these weapons.

d. Automatic Rifles

(1) The platoon commander designates the general fighting positions and principal directions of fire for specific automatic rifles.

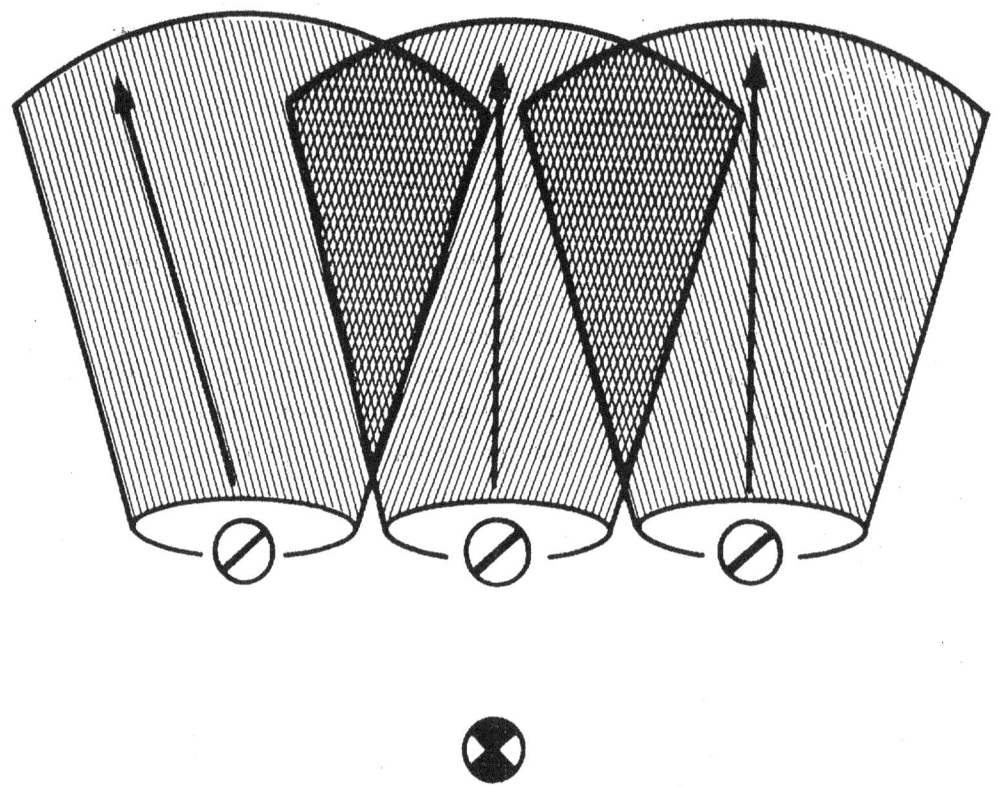

Figure 5-9. Squad Fire Plan.

(2) The squad leader will assign a principal direction of fire for each automatic rifle not assigned by the platoon commander.

(3) The squad leader selects the exact fighting position for each automatic rifle.

e. Position of the Squad Leader. The squad leader's fighting position is usually slightly to the rear of the fire teams and in the center of the squad fighting position. It must be a position from which the squad leader can—

• Observe his squad's assigned sector of fire.

- Observe as much of the squad fighting position as possible, particularly the positions of the fire team leaders.

- Maintain contact with the platoon commander.

f. Fire Plan Sketch. The squad leader prepares the squad fire plan sketch in duplicate. He gives one sketch to the platoon commander for his approval and keeps a copy for himself. The sketch should include fire team fighting positions and sectors of fire, fighting positions and principal directions of fire of the automatic rifles, and the squad leader's fighting position. If the rifle squad is providing protection for a crew-served weapon, its position and primary fire mission (final protective line for machine guns and principal direction of fire for other crew-served weapons) should be included as part of the sketch. Figure 5-10 is an example of a squad fire plan sketch.

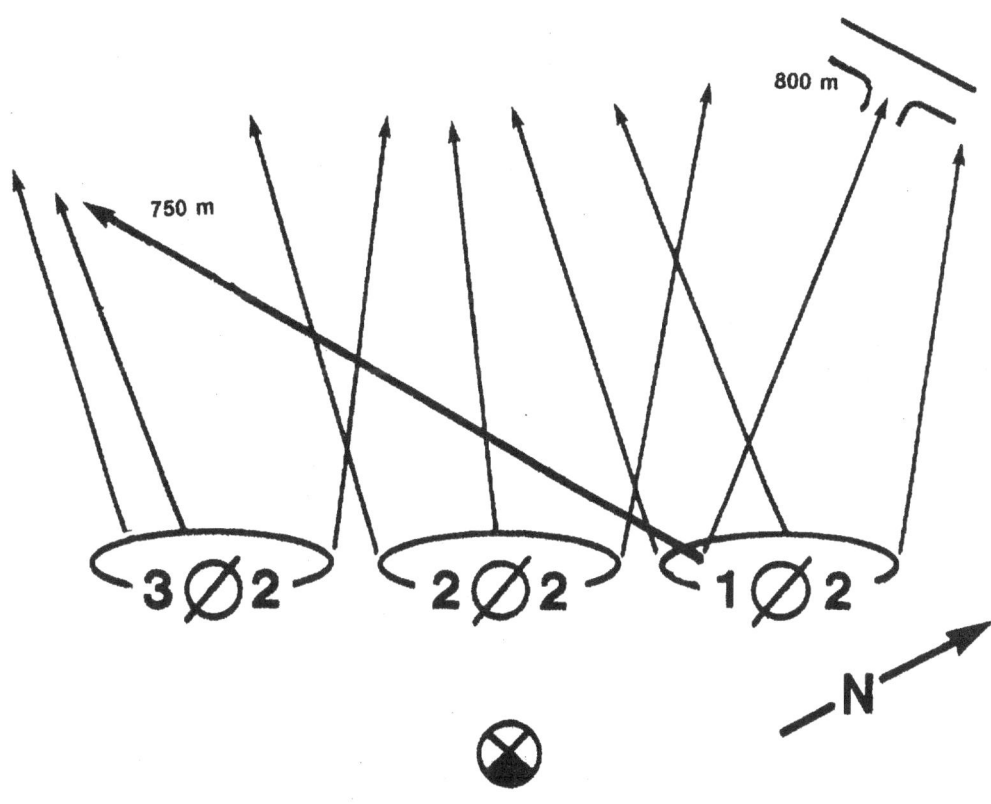

Figure 5-10. Squad Fire Plan Sketch.

Section II. Defensive Procedures

5201. Troop Leading Procedures in the Defense

a. General. Upon receiving the platoon defense order, the squad leader follows the troop leading steps to make the best use of time, equipment, and personnel. Utilizing these steps and satisfactorily completing an estimate of the situation, he issues his squad defensive order. This follows the five-paragraph order format which includes —

- Information about the enemy, the location and identification of adjacent units, and the location of supporting weapons within the squad area.

- The mission of the squad.

- The fighting positions and sectors of fire for each fire team and the principal direction of fire for each automatic rifleman.

- The assignment of light assault weapons to squad members whose fighting positions will cover avenues of approach for armored vehicles.

- Organization of the ground, priority of work, squad security, and any other instructions the squad leader believes necessary.

- The administrative and supply details such as ammunition resupply and the location of medical corpsmen and the aid station.

- Prearranged signals such as pyrotechnics or audible signals, designating when to open fire or deliver final protective fires. The location of the squad leader and platoon commander are also given.

b. Guidance. For troop leading procedures, see appendix C; for estimate of the situation, see appendix D; for the five-paragraph order, see appendix E.

5202. Squad Plan of Defense

After issuing the squad defensive order, the squad leader positions his fire teams to cover the assigned sector of fire. Before detailed preparations of fighting positions are begun, the squad leader verifies the sector of fire of each fire team and the ability of the fire team to observe its assigned sector.

During his check of the fighting positions, he ensures the sectors of fire overlap and that the desired density of fire can be delivered on avenues of approach. The squad leader's responsibilities during the preparation of the fighting position include —

- In conjunction with fire team leaders, inspecting the fighting position for each fire team member, verifying each man's ability to cover the fire team sector of fire.
- Selecting fighting positions for the automatic riflemen, verifying each one's ability to cover the assigned fire team sector of fire.
- Assigning each automatic rifleman a principal direction of fire (PDF) covering a likely avenue of enemy approach, ensuring the PDF is within the sector of fire assigned to the fire team.
- Coordinating with crew-served weapons personnel located in the squad position.
- Supervising the preparation of fighting holes.
- Supervising the clearing of fields of fire.
- Providing security by assigning sentinels or observation posts.
- Coordinating all security measures with adjacent squads and the platoon commander.
- Inspecting fighting positions to ensure that camouflage and overhead cover are satisfactory.
- Supervising the preparation of supplementary fighting positions.
- Establishing a system of signals for fire control.

a. Signal to Commence Firing. Normally, a forward limit is established to designate the range at which the fire teams are to open fire. For rifles and automatic rifles, this may extend as far forward as their maximum effective range. A terrain feature should be selected to locate the forward limit. As the attacker passes this limit, he is brought under fire. This establishes a positive means of fire control to ensure that small arms fire does not commence prematurely or is withheld too long. The squad leader may desire the fire teams to hold their fire until the enemy gets closer than maximum effective small arms range, and then deliver a heavy volume of surprise fire. In this case, he will establish a signal for commencing fire. When the squad commences fire, rifles and M-203s are fired at the average rate; automatic rifles are normally fired at the sustained rate. The squad leader determines what rate of fire is appropriate for the situation. As the enemy comes closer, the rate of fire is increased.

b. Signal to Commence Final Protective Fires. Final protective fires consist of machine gun fires, mortar and artillery fires, automatic rifle and rifle fires, and M-203 fires. The signal to commence these fires is a pre-arranged pyrotechnic or audible signal and is normally passed to the squad from the platoon commander. When this signal is given, the rifles and M-203s continue to fire at the average rate; the automatic riflemen increase their volume of fire to the rapid rate of fire if they have not already reached it. Since the squad has been increasing its rate of fire as the enemy closes, the automatic rifle may already be firing at the rapid rate, or close to it, by the time the signal to fire the final protective fires is given.

c. Signal to Cease Final Protective Fires. Predetermined signals are used to cease final protective fires. When the enemy assault is repulsed, the signal to cease final protective fires is given. When this signal is given, rifles and M-203s may continue to fire at the average rate; the automatic rifles may return to the sustained rate. The rates of fire will be determined by the squad leader and must be sufficient to destroy the enemy remaining to the squad front. The squad leader will determine when it is safe to cease fire entirely.

5203. Squad Security

The rifle squad provides for its own local security by maintaining constant observation to the front, flanks, and rear. Enough men are kept alert at all times to maintain an effective warning system against enemy air and ground activity. In open terrain during daylight, one sentinel per squad is usually sufficient. Under conditions of reduced visibility, one sentinel per fire team is usually assigned. Sentinels should be relieved every two hours, day or night, in order to ensure they remain alert and effective. Prior to posting, they must be briefed on the location and activity of friendly and known enemy forces (including patrols), the password and countersign, the location of the squad leader, and the location of the platoon and company command posts. Sentinels normally man the automatic rifle.

5204. Organization of the Ground

a. General. The organization of the ground begins as soon as individual members of the squad have been assigned sectors of fire. It includes the following tasks:

- Posting security (listening posts, observation posts, patrols).
- Positioning automatic weapons.
- Clearing fields of fire.
- Digging fighting holes.
- Constructing obstacles.
- Selecting supplementary fighting positions.
- Camouflage measures.

b. Posting Security. Local security consists of measures taken to prevent surprise and to deny the enemy information concerning the plan of defense. All-round security and protection against surprise are achieved by—

- Posting a sentinel for surveillance.
- Enforcing noise and light discipline.
- Keeping movement within the squad fighting position to a minimum.

c. Positioning Automatic Rifles. Automatic rifles are positioned to cover the most likely avenues of approach into the squad area. Their positions should enable them to cover the fire team's sector of fire, provide support for adjacent fire teams, and effectively deliver final protective fires.

d. Clearing Fields of Fire. In clearing fields of fire forward of each fighting position, the following guidelines should be observed:

(1) Do not disclose the squad's fighting position by excessive or careless clearing. (See fig. 5-11.)

(2) Start clearing near the fighting position and work forward to the limits of effective small arms fire.

(3) In all cases, leave a thin natural screen of foliage to hide fighting positions.

(4) In sparsely wooded areas, remove the lower branches of scattered large trees. It may be desirable to remove entire trees which might be used as reference points for enemy fire.

Figure 5-11. Clearing Fields of Fire.

(5) In heavy woods, complete clearing of the field of fire is neither possible or desirable. Restrict work to thinning undergrowth and removing lower branches of large trees. In addition, clear narrow lanes of fire for automatic weapons.

(6) If practical, demolish buildings and walls forward of the fighting position which may obstruct fields of fire or provide cover and concealment to the enemy.

(7) Move cut brush to locations where it will not furnish concealment to the enemy or disclose the squad's fighting position.

(8) Extreme care must be taken by the fire team leader to ensure that fields of fire are cleared of obstructions which might cause premature detonation of the M203 projectile.

e. Digging Fighting Holes. Fighting holes are dug by Marines at their fighting positions. Fighting holes provide excellent protection against small arms fire, shell fragments, airplane strafings or bombings, the effects of nuclear detonations, and the crushing action of tanks. If not prescribed by higher authority, the squad leader will designate either one- or two-man fighting holes. The type of fighting hole used is based upon squad strength, fields of fire, size of squad sector of fire, and morale. However, the two-man fighting hole permits one Marine to rest while the other maintains security over the assigned frontage.

(1) **One-Man Fighting Hole**

(a) **Dimensions.** The size and shape of the fighting hole are affected by certain important considerations. It is as small as practicable, exposing a minimum target to enemy fire; wide enough to accommodate the shoulders of a man sitting on the fire step; long enough to permit the use of an entrenching tool; and at least 4 feet deep to the fire step. Standing on the fire step, the Marine should be able to aim and fire his weapon.

1 **Water Sump.** A water sump, below the fire step, is dug at one side of the fighting hole to collect water and provide a space for the Marine's feet while he's seated on the firing step. (See figs. 5-12 and 5-13.)

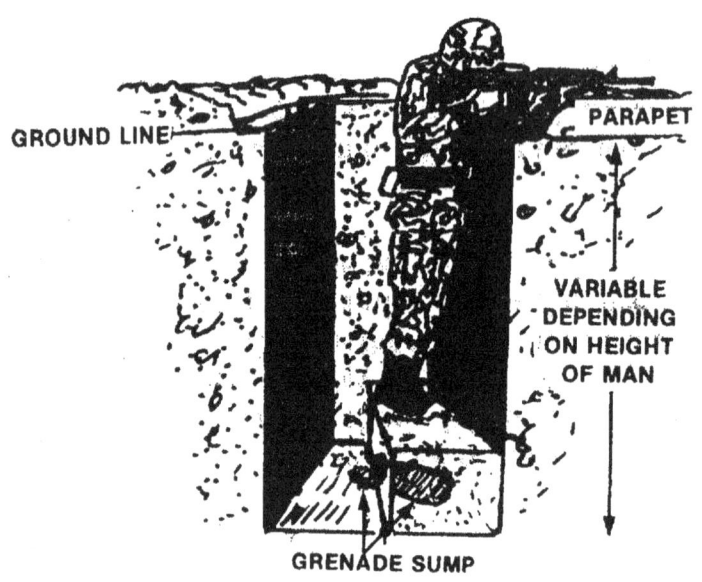

Figure 5-12. One-Man Fighting Hole.
(Horizontal View)

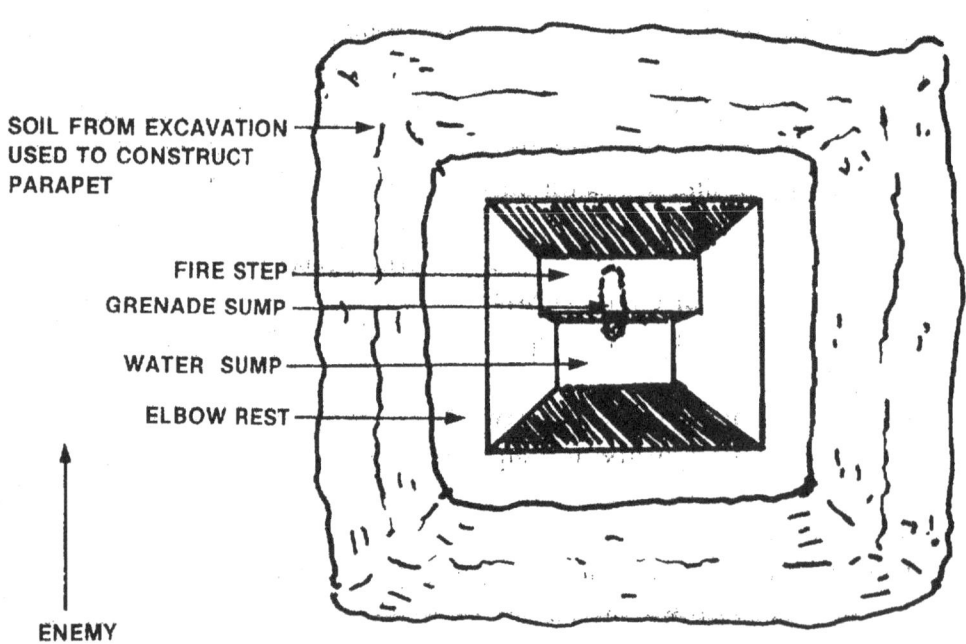

Figure 5-13. One-Man Fighting Hole.
(Vertical View)

__2__ **Grenade Sump.** A circular grenade sump is dug into the wall of the fighting hole facing the enemy, at the lower part of the water sump. The grenade sump should be cone-shaped, with the opening measuring approximately as wide as the spade of the entrenching tool, narrowing to about five inches in diameter at the end; it should be sloped downward at an angle of 30 degrees; and it should be as deep as the Marine can make it. (See fig. 5-12.)

(b) Details of Construction. In most types of soil the fighting hole gives protection against the crushing action of tanks, provided the occupant crouches at least 2 feet below the ground surface. (See fig. 5-14.) In sandy or soft soils it is necessary to revet the sides to prevent caving in. The soil is piled around the hole as a parapet, approximately 3-feet thick and ½-foot high, leaving a berm or shelf wide enough for the Marine to use as an elbow rest while firing. If turf or topsoil is used to camouflage the parapet, the Marine first removes sufficient ground cover and sets it aside until the fighting hole is completed. Once complete, the ground cover can then be laid on the top and sides of the parapet, so that it will better blend in with the surrounding ground.

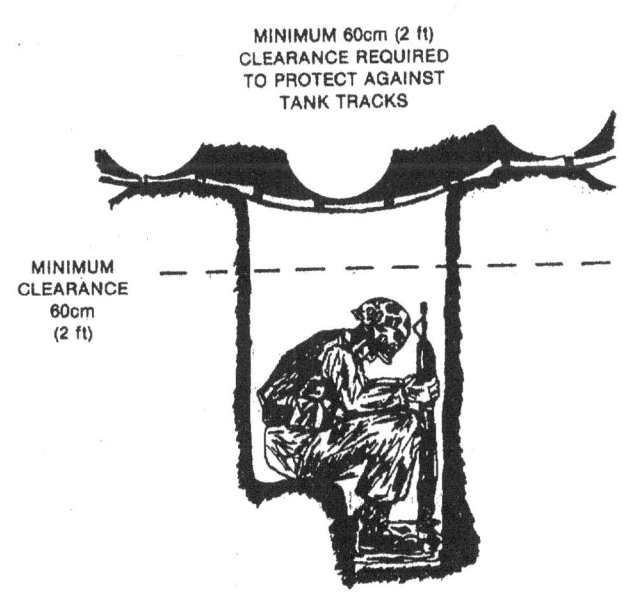

Figure 5-14. One-Man Fighting Hole Protects Against Tanks.

(c) Fighting Hole With Camouflaged Overhead Cover. It is desirable that the soil be removed to an inconspicuous place and a camouflaged overhead cover be improvised. Branches, supporting sod, or other natural material in the vicinity may be used for this purpose. The overhead cover may be reinforced to provide protection from overhead bursts of artillery fire.

(2) Two-Man Fighting Hole. The two-man fighting hole consists essentially of two adjacent one-man fighting holes. Since it is longer than the one-man type, the two-man fighting hole offers somewhat less protection against a tank crossing along the long axis, as well as less protection against strafing, bombing, and shell fragments. (See fig. 5-15.) Some advantages of the two-man fighting hole are that it allows continuous observation, mutual assistance and reassurance, and the redistribution of ammunition between the occupants.

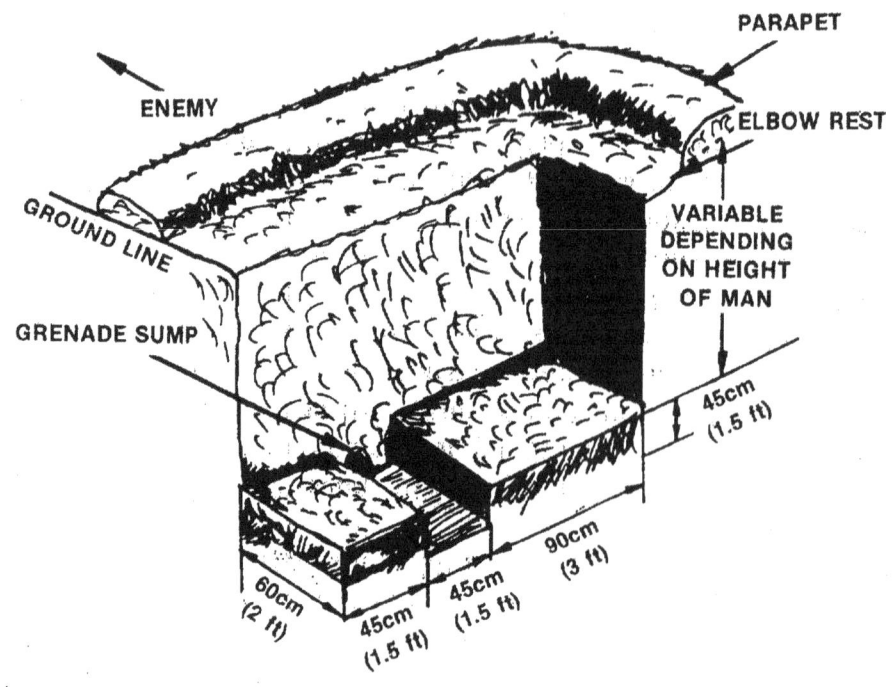

Figure 5-15. Two-Man Fighting Hole.

f. Constructing Obstacles. The squad may be ordered to construct obstacles such as barbed wire, log and brush barriers, ditches, and hasty protective minefields; and may be ordered to improve natural obstacles such as creek beds and river banks. Usually antitank and other extensive obstacles are constructed by engineers. When obstacles which affect the squad's fighting position are constructed, the squad leader ensures that—

- The obstacle is located beyond hand grenade range of the individual fighting positions of squad members.

- The obstacle is covered by fire.

g. Selecting Supplementary and Alternate Fighting Positions. The squad prepares supplementary fighting positions organized the same as the primary fighting positions but oriented in a different direction. If crew-served weapons are attached to the squad or employed in the squad fighting position, alternate fighting positions should also be prepared for the crew-served weapons.

h. Camouflage Measures. Concealment from enemy ground and aerial observation is very important in selecting and organizing each fighting position. The squad must take advantage of natural concealment whenever possible. Camouflage measures are begun from the moment the position is occupied and are continued as long as the Marines are there. Specific camouflage measures are:

(1) Do not disclose the position by excessive or careless clearing of fields of fire.

(2) Use the same turf or topsoil that has been removed from the area of the fighting hole to camouflage the parapet.

(3) Dispose of all soil from the fighting hole not used on the parapet. Carry the soil away in sandbags or shelter halves. Dispose of it under low bushes, on dirt roads or paths, in streams or ponds, or camouflage it.

(4) Avoid digging in next to an isolated bush, tree, or clump of vegetation.

(5) Conceal the fighting hole from observation by the use of a camouflaged cover. Construct the cover from natural materials.

(6) Replace natural material used in camouflage before it wilts or changes color.

(7) Avoid creating fresh paths near the position. Use old paths or vary the route followed to and from the position.

(8) Avoid littering the area near the position with paper, tin cans, and other debris.

5205. Squad Defense Order

The squad leader follows the standard five-paragraph order format in presenting his squad defense order. (See app. E.) A sample defense order given by the squad leader, 3d squad, commencing with a terrain orientation, follows:

(Terrain Orientation) "That direction is north (pointing). Notice the stream bed to the front, that road on the left, that destroyed bridge, and the woods on the left.

(Paragraph 1) "An enemy force supported by tanks, artillery, and aircraft is expected to attack from that direction (pointing), sometime after midnight tonight.

"Our platoon will defend this high ground from just this side of the road (pointing) to a point 500 meters to the right (pointing). Our fighting position runs along the forward slope of the high ground (pointing).

"The 2d squad is on our right and the 2d platoon on our left.

"There is a machine gun squad in the 2d platoon area that fires to the right, across our front, and another in the 1st squad area that fires to the left, across our front. Two SMAW teams are located in our area just to the right of the road and fire down the road. Mortar final protective fires will fall in the stream bed to our front and an artillery final protective fire will fall in the vicinity of the road.

"A platoon-size security force now in position to our front will withdraw along the road, probably sometime tonight, if the expected enemy contact proves too strong. I'll get the word to you as soon as I find out. Make sure all your men know they're out there.

(Paragraph 2) "The mission of our squad is to organize and defend a part of the platoon battle position from the right side of the road over to and around this finger, over to and including the draw to the right. Our sector of fire is the area between that bend in the stream on our right (pointing) and the break in the woods on our left (pointing).

(Paragraph 3) "Our squad will organize our defense with three fire teams on line. One automatic rifle's principal direction of fire will be down the drainage ditch along the right side of the road. One automatic rifle PDF will fire down the path in the center of the finger. The PDF of the other automatic rifle will be down the draw to the right.

"1st fire team, on the right, will defend from that draw (pointing), around the right side of the finger to, and including, that tree stump. Your sector of fire will extend from that bend in the stream in second squad's area (pointing), left to the other side of that large rock (pointing). Your automatic rifleman will fire his PDF down the draw. I want it to be fired from this position (pointing) to that old dead tree there (pointing).

"2d fire team, in the center, will defend from that tree stump, to and including, that bush (pointing) at the left center of the finger. Your sector of fire is from the demolished bridge on the left to that clump of cattails there in the stream bed (pointing). Place your automatic rifleman here and have him fire his principal direction of fire down the path running the center of the finger. Your fire team will post one man to act as security for the squad while we are digging in. Have him remain on this high ground, and have him watch that stream bed in particular. Give him the automatic rifle. I will have him relieved in one hour.

"3d fire team, on the left, will defend from that bush at the left center of the finger to the road. The road itself is in 2d platoon's area. Your sector of fire is from that large rock in the stream bed (pointing) to that large tree in 2d platoon's area (pointing). The PDF for your automatic rifle is down the drainage ditch on the right side of the road.

"I will point out supplementary fighting positions protecting the rear later.

"After I have checked each man's fighting position and his coverage of the fire team's sector of fire, we'll clear fields of fire, dig one-man fighting holes with overhead cover, and camouflage them at the same time. Fire team leaders assign tasks. I have already coordinated the overlap of sectors of fire with 2d squad on the right and 2d platoon on the left.

"Open fire on the enemy when they come out of the woods to our front. We will use the woods as the forward limit.

(Paragraph 4) "Water and rations will be issued before sunset. Make sure all hands have four grenades.

"The battalion aid station is along that road about 800 meters to the rear. The platoon corpsman is near the platoon CP over there (pointing).

"Send POWs back to me.

(Paragraph 5) "The challenge is 'September' and the password is 'Beacon'.

"Signal to commence firing the final protective fires is a red star cluster. Signal to cease firing the final protective fires is a green star cluster.

"The platoon commander is in the edge of the woods to our right rear (pointing).

"My position will be here on this finger just behind the 2d fire team.

"Any questions?

"It is now 1400.

"Move out!"

5206. Conduct of the Defense

a. **Enemy Preparatory Bombardment.** The enemy will normally precede his attack with fire from any or all of the following weapons; artillery, naval gunfire, mortars, machine guns, tanks, and aircraft. During this incoming enemy fire, the squad will take cover in its fighting holes, maintaining surveillance to the front, flanks, and rear to determine if the enemy is advancing closely behind their supporting fires.

b. **Opening Fire and Fire Control.** The squad withholds its fire on approaching enemy troops until they come within effective small arms range of the squad's fighting position. Squad members open fire on the approaching enemy on command of the squad leader, or when the enemy reaches a predetermined line, normally the forward limit of the fire team sector of fire. When the squad opens fire, rifles are fired at the average rate. When the enemy enters the range of the M-203, the fire team leader delivers grenade launcher fire at the average rate. Automatic riflemen normally fire at the sustained rate. The squad leader determines the appropriate rate of fire for the situation. Automatic riflemen's priority of fire goes to enemy automatic weapons, rocket launchers, and other crew-served weapons. Once the squad opens fire, direct control passes to the fire team leaders. The fire team leaders, in accordance with the squad leader's previous plan, designate new targets, change rates of fire when necessary, and give the order to cease fire when the attack is defeated. The goal of the squad is to defeat the enemy attack as far forward of the squad fighting position as possible. If the enemy is not stopped and he continues to close on the squad fighting position, the automatic riflemen will continue to increase their rate of fire as the enemy comes closer.

c. Final Protective Fires. If the enemy's attack is not broken and he begins his assault, final protective fires are called. Final protective fires are the final attempt to stop the enemy attack before he reaches the platoon's battle position. When final protective fires are called for, all squad members fire in their assigned sectors (normally the fire team's sector of fire). Rifles and M-203s continue to fire at the average rate; the automatic riflemen will increase their volume of fire to the rapid rate, if they have not yet reached this rate prior to the calling for final protective fires. Riflemen engage enemy personnel within the fire team sector; fire team leaders fire the M-203 at the largest concentration of enemy personnel within the fire team sector. Normally, the largest concentrations will be along the PDFs of the automatic rifles if the PDFs were properly positioned.

d. Enemy Reaches the Fighting Squad's Position. Enemy infantry reaching the squad's fighting position are driven out by fire, grenades, the bayonet, and hand-to-hand combat. The success of the defense depends upon each rifle squad defending in place. A stubborn defense by front line squads breaks up enemy attack formations and makes him vulnerable to counterattack by reserve units. The squad does not withdraw except when specifically directed by higher authority.

5207. Defense Against Mechanized Attack

When tanks or other armored vehicles support an enemy infantry attack, the primary target of the squad is the hostile infantry. This holds true whether the enemy infantry is on foot (dismounted), mounted in armored personnel carriers (APCs) or in trucks. If the enemy infantry is mounted in trucks, they can be engaged with small arms; if in APCs, they can be engaged with small arms using armor-piercing ammunition (if available) and light assault weapons (LAWs). The goal is to slow down the infantry movement by making them dismount. This will either separate the enemy infantry from the tanks or force the tanks to slow down to keep pace with the dismounted infantry. When hostile infantry does not afford a target, the squad may direct its small arms fire and LAWs against the aiming devices and vision slits of enemy armor. LAWs are used to destroy enemy tanks or to damage the tracks and suspension system to the point where the tank can no longer move (mobility kill). Under no circumstances will the squad be diverted from its basic mission of engaging and destroying the hostile infantry. Every effort is made to separate the enemy tanks from the enemy infantry because the tanks, even if they pass through the squad defensive position, are very vulnerable to crew-served antitank weapons once they are stripped of the supporting infantry.

5208. Movement to Supplementary Fighting Positions

If the fighting position of an adjacent squad is penetrated by the enemy, the squad leader shifts a part of the squad's fire into the penetrated area, and, if necessary, moves some men to supplementary fighting positions protecting the threatened flank. If the squad fighting position is threatened by attack from the rear, the squad leader moves some men to supplementary fighting positions protecting the rear. In open flat terrain, the squad leader simply orders his men to shift their fire to the rear. Prior to moving men to supplementary fighting positions, the squad leader, if possible, requests the approval of the platoon commander. When it is not possible to request permission, the squad leader notifies the platoon commander of his action as soon as possible. The squad leader avoids moving an entire fire team to supplementary fighting positions, but instead moves one or two men from each fire team, depending on the number required to protect the flank or rear. In any case, men moving to supplementary fighting positions follow the route which affords the best cover.

5209. Local Security for Platoons and Companies

The squad often furnishes local security for the platoon and company. Security posts from two to four men are stationed by the platoon commander or company commander up to 460 meters (effective small arms range) forward of the platoon battle position. Small patrols are often used to cover the ground between security posts or as a substitute for security posts. The company commander or platoon commander designates the general positions to be occupied by the security posts and the routes to be covered by the patrols. The squad leader may find his squad divided into small security posts and patrols covering the platoon or company front and flanks. His duties then include—

- Checking to see that security posts are well concealed and permit observation of the ground over which the enemy is expected to advance.

- Checking to see that patrols are following the prescribed routes.

- Passing on to his men all available information regarding both friendly and enemy forces.

- Instructing his men as to what action to take in case of enemy attack.

- Informing the platoon commander or company commander immediately of enemy activity.

When the enemy approaches, security posts and patrols take the following actions:

- Notify the platoon commander or company commander immediately of the enemy's strength, actions, direction of advance, and weapons and equipment.

- On order, withdraw along a predetermined route to the platoon battle position in sufficient time to prevent being engaged in close combat. After reaching the platoon battle position, report all information regarding the enemy to the commander who originally ordered the patrol or security element.

5210. Security Forces (Formerly, Combat Outpost)

  a. The squad may serve as part of a security force. Security forces are assigned three types of missions— screen, guard, and cover.

(1) A **screen** is a security element whose primary task is to observe and report information, avoiding decisive engagement with the enemy. A screen accomplishes the following tasks:

- Provides early warning of enemy approach.

- Gains and maintains enemy contact and reports enemy activity.

- Within capabilities, destroys or repels enemy reconnaissance units.

- Impedes and harasses the enemy with indirect fires.

- Guides reaction forces.

(2) A **guard** protects the main force from attack, direct fire, and ground observation by fighting to gain time, while also observing and reporting information. A guard accomplishes the following tasks:

- Provides early warning and maneuver space to the front, flanks, and rear of the main force.

- Attacks, defends, or delays, within its capabilities, to protect the main force.

(3) A **covering force** is a force which operates apart from the main force for the purpose of intercepting, engaging, delaying, and deceiving the enemy before he can attack the main force. A covering force accomplishes the following tasks:

- Gains contact with the enemy.

- Protects the main force from engagement.

- Denies the enemy information about the size, strength, composition, and objective of the main force.

- Destroys enemy reconnaissance and security forces.

- Develops the situation to determine enemy disposition, strengths, and weaknesses.

b. The location and composition of the security force is determined by the commander of the main force. He will organize the security force according to the mission he gives it—screen, guard, or cover.

c. Generally, the role of the rifle squad as part of a security force will be the same, regardless of the mission assigned to the security force. The squad will report enemy sightings, take the enemy under fire, and withdraw only on orders from the platoon commander or the commander of the security force.

d. Withdrawal routes will have been previously determined and reconnoitered. Upon withdrawal and passage through the forward friendly unit, the squad will return to its parent platoon (if the squad had been operating independent of the platoon), which is normally part of a reserve unit.

Chapter 6

Amphibious Operations

6001. Introduction

The purpose of an amphibious operation is to launch an attack from the sea on a hostile shore.

6002. Preembarkation

Preembarkation encompasses those functions that must be performed to prepare the squad for an amphibious operation. The following guidelines are to be used by the squad leader in preparation for embarking aboard ship:

- Supervise individual and unit training.

- Ensure fire team leaders are proficient in fire team and squad tactics.

- Conduct critiques of squad training and initiate corrective action.

- Conduct inspections of weapons, clothing, and equipment to ensure readiness for embarkation and the operation.

- Supervise the marking/tagging of weapons, equipment, and baggage for embarkation.

- Assemble squad in proper uniform at designated time for embarkation.

6003. Duties Aboard Ship

The squad leader performs the following duties while the ship is underway:

a. Assigns fire team berthing areas and supervises the stowage of gear.

b. Ensures squad attends all briefings, ship's drills, and periods of instruction.

c. Ensures squad area is in good state of police.

d. Enforces applicable ships' regulations.

e. Assists in conducting and supervising physical conditioning and military subjects' training.

f. Directs and supervises care and cleaning of weapons.

g. Supervises security of individual weapons.

h. Ensures correct conduct and appearance of squad.

i. Conducts operational planning.

  (1) Briefs squad on:

- Platoon mission.
- Scheduled rehearsals.
- Debarkation procedure.
- Ship-to-shore movement.

  (2) Makes a detailed study of:

- Maps.
- Aerial photographs.
- Mockups and sketches.

  (3) Makes a preliminary estimate of the situation.

  (4) Formulates tentative plan of attack.

  (5) Submits tentative plan of attack to platoon commander.

  (6) Completes tentative plan.

j. Issues the order and ensures thorough understanding by all members of the squad.

k. Additional duties may be prescribed for the squad leader and members of his squad.

6004. Debarkation

Debarkation is characterized by rapid and effective unloading of men and material in the shortest possible time.

a. Organization of a Boat Team. A boat team includes the personnel, equipment, and supplies assigned to one landing craft or amphibious vehicle. They are normally organized as follows:

(1) Landing Craft

- Boat team commander.
- Assistant boat team commander.
- Four net handlers.
- Boat paddle handler.
- Eight loaders (four deck loaders and four boat loaders).
- Remaining troops and equipment.

(2) Assault Amphibious Vehicle (AAV)

- Boat team commander.
- Assistant boat team commander.
- Remaining troops and equipment.

b. Duties of Boat Team Personnel

(1) Boat Team Commander. The senior commissioned or noncommissioned troop officer of the boat team is designated the boat team commander. He has the following responsibilities:

(a) Appointing the assistant boat team commander, loaders, net handlers, and boat sign handler.

(b) Assigning personnel and equipment to positions in the landing craft or AAV.

(c) Reconnoitering routes from assigned troop assembly area to debarkation station or assigned AAV.

(d) Mustering boat teams in assembly area at the required time.

(e) Inspecting each man for uniform, equipment, and adjustment of equipment while in the assembly area.

(f) Ensuring that lashing lines are present for each piece of equipment that is to be lowered into the landing craft and that all such equipment is properly lashed.

(g) Forming his boat team at the debarkation station (not applicable to AAVs).

(h) Supervising the debarkation of personnel and the lowering of equipment during debarkation (not applicable to AAVs).

(i) Maintaining boat team discipline during the ship-to-shore movement. The boat team commander has no control over the coxswain or boat crew during the ship-to-shore movement.

(j) Debarking boat team personnel and equipment from the landing craft or AAV at the beach.

(2) Assistant Boat Team Commander. The second senior commissioned or noncommissioned troop officer present normally is assigned duty as assistant boat team commander. His duty is to assist the boat team commander in carrying out duties outlined in paragraph (1), above.

(3) Loaders. Eight members of the boat team are designated as loaders. The loaders lower, guide, and stow all equipment which cannot be carried down the debarkation net. Four men are designated deck loaders and lower gear with lowering lines from the ship's deck to the boat. Four boat loaders in the boat guide the equipment away from the side of the ship with guidelines, receive it, and stow it in the boat.

(4) Net Handlers. Four men are normally assigned duty as net handlers, but additional net handlers may be assigned, depending on the condition of the sea in the objective area. The net handlers are the first group to go down the net and relieve the boat crew of holding the debarkation net taut and away from the side of the landing craft.

(5) Boat Paddle Handler. One member of the boat team is responsible for the boat paddle. The boat paddle serves to identify the boat team and aids the wave commander in forming a wave in proper order in the rendezvous area. The handler lashes it and carries it from the troop assembly area to the debarkation station where it is lowered as prescribed for other equipment. When directed by the boat team commander, the handler places the paddle in the boat paddle bracket on the starboard side forward. He removes the paddle after the line of departure is crossed, carries it ashore, and drops it on the beach above the high water mark.

c. Individual Equipment Preparations. Equipment, except for body armor, is rigged so that it may be quickly dropped from the shoulders and discarded in the event that the individual falls into deep water while debarking from the ship. Equipment is rigged on the individual in the boat team assembly area as follows:

(1) Protective mask is slung over the right shoulder (not around the neck), riding on the right hip with the body strap unfastened and wrapped around the cartridge belt.

(2) Canteen well back on the left hip. If two canteens are worn, one is carried on each hip.

(3) This procedure can be performed at the debarkation station. The shoulder weapon is slung on the right shoulder with the muzzle up, sling loosened and to the front. The weapon is carried across the pack with the sling around the pack, bringing the weapon to a vertical position behind the left shoulder. The sling is secured in this position. The weapon should fit snugly. Men pair off to adjust weapons; one man placing the weapon on the other's pack, the other reaching down with his left hand grasping the butt of his weapon to test it for snugness and to hold it in position in order that proper sling adjustment may be made. The sling will be fastened after the weapon has been properly adjusted.

(4) The lanyard of the pistol is placed over the right shoulder (not around the neck) and fastened securely to the pistol.

(5) Suspender straps attached to flak jacket; cartridge or pistol belt is left unfastened.

(6) Helmet chin strap is loose but fastened.

(7) Body armor is unfastened.

(8) Life jacket:

   (a) **Cork or Kapok.** This life jacket is put on after all other individual equipment is adjusted. The jacket is placed around the man's neck; crotch and waist straps are brought under the individual equipment, over the body armor, care being taken not to entangle the straps with any other equipment; crotch and waist straps are drawn tight; the tie-tie straps are tied. This type is not used in AAVs.

   (b) **Pneumatic Type.** This type of life jacket is put on over the body armor before other individual equipment is put on. The back, body, and crotch straps are fastened, leaving sufficient looseness to allow for inflation.

(9) All equipment, except for life jacket, is adjusted so that it can be dropped quickly from the shoulders.

(10) Gloves are removed.

d. Movement From Assembly Area to Debarkation Station. The boat team commander leads his boat team in single file to its assigned debarkation station.

e. Procedure for Debarkation From Transports. Men and equipment should be debarked as quickly as possible when their assigned boats come alongside. The boat team commander is in charge of the debarkation of his team. He orders men over the side in ranks of four. Nets are used to full capacity; the men in each rank keep abreast of each other by glancing to the right and left, allowing the slowest men to set the pace.

f. Lashing and Lowering the Equipment. Each piece of equipment to be lowered is lashed with a line prior to its arrival at the debarkation station. The lash line should be of sufficient strength to hold 300 pounds (not less than ½ inch in diameter) and have a 4-inch eye splice at each end. Lowering lines and guidelines furnished by the ship will be at each debarkation station. The guideline is hooked in the eye of the lashing line at the heavy end of the equipment and the lowering line is hooked at the light end. In lowering, the guideline is kept taut, keeping the equipment away from the side of the ship and guiding it into the landing craft. Some amphibious ships provide canvas bags, which can be used in place of lashing lines, for use in lowering equipment.

g. AAV Boat Team. Organization and composition of the AAV boat team is given in paragraph 6004a(2). Preparation for debarkation, as given in paragraphs 6004b and 6004c, applies to AAVs with the exception that unit equipment is preloaded and properly stowed in the AAV and shoulder weapons are slung normally or carried in the hand.

6005. Movement From Ship To Shore

a. Duties Aboard Landing Craft. When the landing craft departs the ship for designated positions in the landing area, the boat team commander performs the following:

(1) Checks to see that men and equipment are in assigned boat spaces.

(2) Ensures the boat paddle is positioned on the starboard side forward.

(3) Upon crossing the line of departure, ensures boat paddle is removed, orders protective covers removed from weapons and equipment, unlashes equipment, fastens all cartridge belts and chin straps, and locks and loads weapons.

(4) Conducts a visual reconnaissance of the beach from the landing craft, observing enemy installations, beach condition, and objective.

b. Duties Aboard AAVs. Duties aboard an AAV are the same as those listed in paragraph 6005a, except for reference to the boat paddle.

c. Debarkation From Landing Craft. Personnel debark on order when the landing craft beaches. The right column moves down the starboard side of the ramp, steps off the right side, and deploys to the right front. The left column moves down the port side of the ramp, steps off the left side, and deploys to the left front. The center column alternates, half debarking over the starboard and half over the port side of the ramp. Life jackets are removed when personnel reach a good covered position.

d. Debarkation From AAVs. Squads debark as a unit to maintain tactical unity. When units reach a good covered position life preservers are removed.

e. Emergency Abandonment From Landing Craft and AAVs

(1) **Landing Craft.** If a landing craft is in danger of sinking, combat equipment is discarded by the individual. Leaders must set the example

and maintain control of members of the boat team. When the landing craft has sunk beneath the boat team, personnel stay together and swim toward the nearest refuge. If the boat is afire, members of the boat team, on order, debark over the side quickly.

(2) **AAV.** The primary principle in escape operations from amphibious vehicles is strict attention to the orders of the vehicle crew chief and maintenance of discipline among the boat team.

(a) Debarkation From a Sinking AAV. On order of the vehicle crew chief, the boat team will jettison equipment, move topside, inflate life jackets, and abandon the vehicle.

(b) Debarkation From a Sunken AAV. In the event of rapid sinking, it may not be possible to evacuate personnel before the vehicle slips under the surface of the water. Some air will be trapped within the personnel and cargo compartment of the sunken AAV, and will serve as a pocket of air when the vehicle has settled. This pocket of air should be sought by all personnel aboard until internal and external pressures have equalized, that is, when the water level in the troop compartment has risen to a level equal to the top of the rear personnel hatch. When equalization has taken place, the amphibious vehicle crew members will open the personnel hatch or vehicle ramp. No attempt should be made to release the hatch until they are totally submerged. Boat team personnel are then led to the hatch by crew members. Life preservers of the inflatable type will not be inflated until the hatch has been cleared.

6006. Amphibious Assault

a. Action Upon Landing. The first priority for the assault wave is the destruction of the enemy on the beach. Each squad is assigned a specific objective. The squad seizes its objective as rapidly and aggressively as possible, regardless of the progress of other elements of the platoon. Whether attacking a beach objective or pressing inland, the action of the squad must be rapid and decisive. Prompt followup of the preliminary bombardment will give the best results. Installations and terrain features may often be seized quickly by immediate attack, whereas, a delay may require a later and more difficult attack.

b. Support Units. Machine gun, SMAW, and mortar squads are normally boated with the assault rifle squads to form the boat teams of the leading wave. These personnel are assigned definite objectives against which they will deliver supporting fire. The squad leader must know how to employ these weapons, and the entire squad must be trained to operate with them.

Chapter 7

Helicopterborne Operations

7001. Introduction

The helicopter provides increased battlefield mobility for ground units. Squad and fire team leaders must be familiar with basic information concerning principles and techniques used in helicopterborne operations.

7002. Concept of Employment

Infantry units can be organized and equipped to conduct helicopterborne operations. Helicopterborne forces can be employed to seize key terrain; to isolate pockets of resistance; to conduct linkups with other forces; and to conduct diversionary actions, raids, combat patrols, deep reconnaissance, observation and surveillance, and operations against guerrillas.

7003. Basic Definitions

a. Helicopter Team. A helicopter team is commonly called a heliteam. A heliteam is the tactical unit, equipment, and supplies lifted in one helicopter at one time. Each heliteam is identified by an assigned serial number, which also identifies that aircraft with its helicopter flight or wave. In forming heliteams, tactical integrity is preserved to the maximun extent possible. The size of the heliteam is determined by the tactical mission, the weight carrying capacity of the helicopter, and the weight of the troops and equipment to be transported. For planning purposes, the weight of a combat-loaded Marine is 240 pounds. The senior man in the heliteam is designated the heliteam leader and is placed in charge of the heliteam.

b. Helicopter Flight/Wave. A helicopter flight consists of the helicopters which arrive together and land at approximately the same time in the same landing zone. During amphibious operations, helicopter flights are called helicopter waves. Normally, a rifle platoon is the smallest tactical unit transported in a single wave.

c. Landing Zone. A landing zone is a specified ground area for landing helicopters to embark or disembark troops and/or cargo. A landing zone is designated by a code name. It may include one or more landing sites. (See fig. 7-1.)

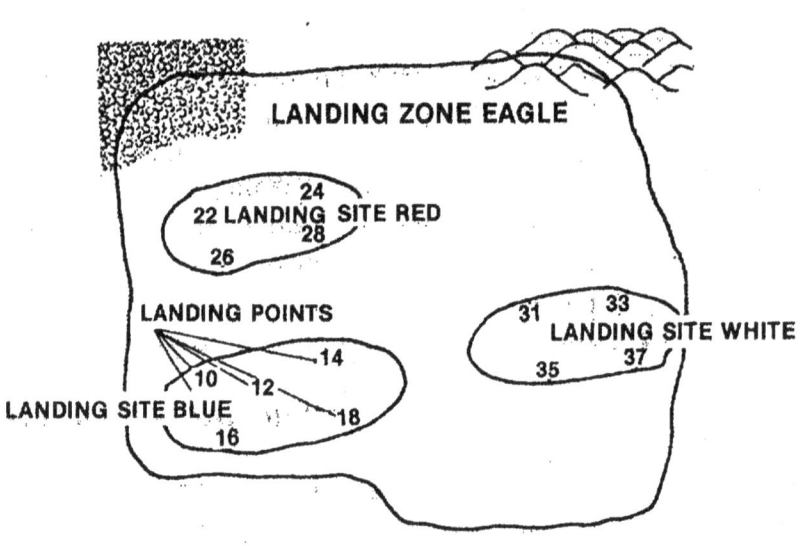

Figure 7-1. Helicopter Landing Zone.

d. Landing Site. A landing site is a specific location within a landing zone, in which a single flight of helicopters may land to embark or disembark troops and/or cargo. Landing sites are designated by a color, such as *landing site red*. A landing site contains one or more landing points.

e. Landing Point. A landing point is a point within a landing site where one vertical assault aircraft can land. Landing points are designated by two-digit numbers, such as *landing point 12*. (See fig. 7-1.)

7004. Helicopterborne Operations Training

The objective of helicopterborne operations training is to familiarize personnel with its techniques and ensure that these operations are conducted with maximum speed, flexibility, and timeliness.

a. Heliteam Organization. Heliteams are organized as follows:

- Heliteam leader.
- Assistant heliteam leader.
- Remaining members of the heliteam and their equipment.

b. Heliteam Leader's Responsibility. The senior commissioned or noncommissioned troop officer of the heliteam is the heliteam leader. His responsibilities are as follows:

- Inspect each individual for proper uniform, equipment, and proper adjustment of equipment while in the assembly area.
- Muster and organize the heliteam in the assigned assembly area.
- Ensure that equipment assigned the heliteam is properly located before the team is called to the holding area or pickup zone.
- Lead his heliteam from the assembly area to the holding area, pickup zone, and loading point.
- Supervise enplaning of his heliteam.
- Supervise deplaning of heliteam personnel and equipment at the landing site.

c. Assistant Heliteam Leader's Responsibility. The second commissioned or noncommissioned troop officer is the assistant heliteam leader. He assists the heliteam leader. He must be thoroughly familiar with all duties of the heliteam leader and assumes leadership when necessary.

d. Loading Procedure. Troops will be assembled into an assembly area. Here orders are issued and administrative matters are completed. Troops are assembled into heliteams and briefed. When directed, the heliteams are moved to the holding area. The assembly area may also serve as the holding area. From the holding area, heliteams are moved to a pickup point within the pickup zone. (See fig. 7-2.)

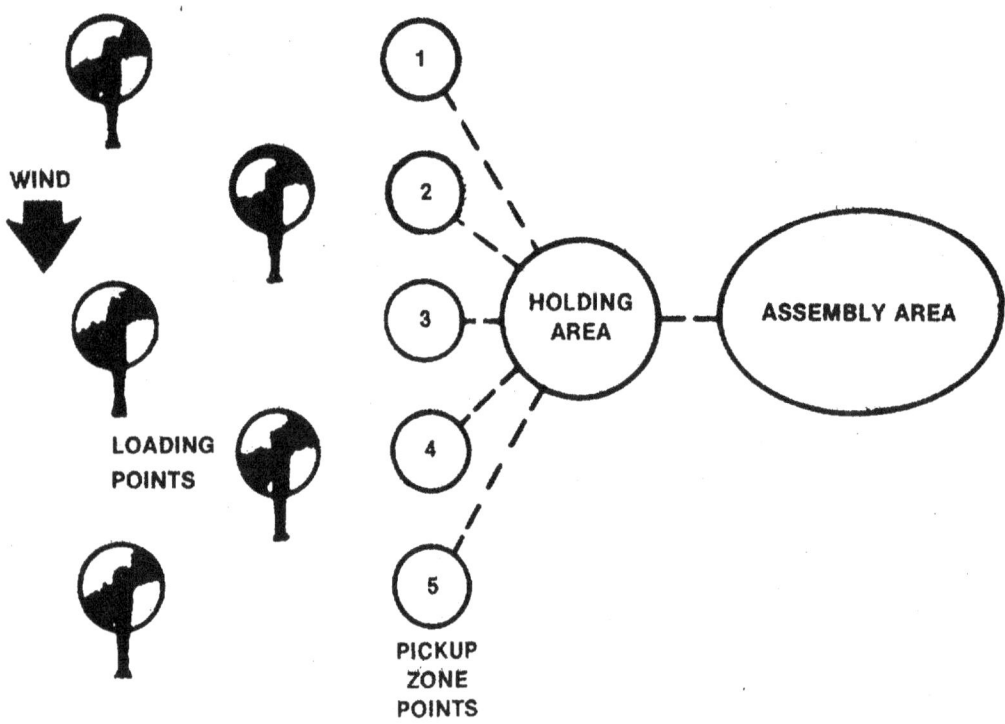

Figure 7-2. Pickup Zone Operation.

e. Loading. Loading is conducted with the maximum speed commensurate with safety. Specific procedures for loading, by type of aircraft and according to the situation, will be prescribed in local SOPs. To assist in loading drills, the following procedures may be used as guides:

(1) On signal from the pickup zone control officer or NCO in the pickup zone, the heliteam approaches the aircraft on the double with the heliteam leader leading and the assistant heliteam leader bringing up the rear.

(2) The heliteam leader ensures that team members are in proper sequence within the column to facilitate rapid loading of personnel and equipment.

(3) Upon reaching the entrance to the aircraft, the heliteam leader takes a position outside, slings his rifle, and assists team members in loading.

(4) Personnel enter the aircraft carrying rifles in their hands.

(5) The passenger manifest is passed from the heliteam leader to the troop loading assistant within the pickup zone control team. See FMFM 6-21, *Tactical Fundamentals of Helicopterborne Operations*, for details concerning passenger manifesting.

(6) When seated, each Marine places his rifle between his knees, fastens his safety belt, and raises his right arm to signal the heliteam leader that he is ready for takeoff.

(7) When the heliteam is strapped in and ready for takeoff, the heliteam leader gives the crew chief a ready signal which has previously been agreed upon by the lifted troop unit and the helicopter unit. Some aircraft have been modified to provide voice radio communications between the pilot and the heliteam leader.

f. Loading Aboard Ship. Loading aboard ship is the same as that ashore. The organization and physical appearance of amphibious assault ships differ according to the ship's class, but general heliteam loading procedures remain the same.

g. Landing. When approaching the landing zone, but shortly before touchdown, the pilot or crew chief will orient the heliteam leader with relation to direction. Besides identifying north, south, east, and west, he must establish for the leader his position in relation to an object that is well known to him on the ground. When the aircraft has landed, the pilot, co-pilot, or the crew chief will give the signal to unload. Personnel will unfasten safety belts and unload quickly. This is necessary to ensure that aircraft are not exposed for a long period in the landing zone.

7005. Conduct of the Helicopterborne Assault

a. Planning. The principles of offensive combat are the same for the employment of helicopterborne forces as for normal land combat. The squad leader prepares for the helicopterborne assault by carrying out the following duties:

(1) Makes a preliminary estimate of the situation.

(2) Conducts map and aerial photograph reconnaissance.

(3) Coordinates with adjacent squad leaders.

(4) Formulates a tentative plan of attack to include:

- Heliteam organization.
- Scheme of maneuver (clearing squad sector of landing site and seizure of objectives).
- Fire support.

(5) Submits his tentative plan of attack to platoon commander.

(6) Briefs the members of the squad on the platoon mission.

(7) Completes the plan, issues the order, and supervises.

b. Initial Ground Action

(1) Initial Assault. The initial assault involves seizing and establishing landing sites. Each squad of the first wave will be assigned a sector of responsibility for all or part of the landing site. Upon landing, the squads will seize their sectors.

(2) Seizure of the Landing Site. When the area is seized, the squad sets up a hasty defense to ensure initial landing zone security. Concurrent with operations, the squad leader accomplishes reorganization and control. Contact with friendly units is established as rapidly as possible.

(3) Followup. Units landed in succeeding waves are employed, as necessary, to ensure seizure of the landing zone.

Chapter 8

Patrolling

Section I. Patrol Organization

8101. General

A patrol is a detachment of ground forces sent out by a larger unit for the purpose of gathering information or carrying out a destructive, harassing, or security mission. Patrols may range in size from fire team to platoon, depending on the type patrol, its mission, and its distance from the parent unit. While most combat patrols should be platoon-sized, reinforced with crew-served weapons, the Marine rifle squad is ideally suited for reconnaissance patrols. For more detailed information on patrolling, see FMFM 6-7, *Scouting and Patrolling*.

8102. General Organization

The platoon commander designates a patrol leader, who is normally one of his squad leaders, and gives him a mission. The patrol leader then establishes his patrol headquarters and the patrol units required to accomplish the mission.

a. Patrol Headquarters. The patrol headquarters is composed of the patrol leader and personnel who provide support for the entire patrol, such as a forward observer, corpsman, and radio operator.

b. Patrol Units. Patrol units are subdivisions of patrols. Personnel are assigned to units based on the mission of the patrol and the mission of individuals within the patrol.

c. Reconnaissance Patrol. A reconnaissance patrol collects information about the enemy, terrain, or resources. It relies on stealth and fights only when necessary to accomplish the mission or defend itself. A reconnaissance patrol is organized into two units.

(1) The reconnaissance unit reconnoiters or maintains surveillance over the objective.

(2) The security unit secures the objective rally point, gives early warning of enemy approach, and protects the reconnaissance unit.

d. Combat Patrol. A combat patrol is a fighting patrol. Because the patrol is assigned a mission which may require it to engage the enemy, a combat patrol is stronger and more heavily armed than a reconnaissance patrol. A combat patrol is assigned a mission to destroy enemy troops, equipment, or installations; capture enemy documents, equipment, or installations; and, as a secondary responsibility, gather information. A combat patrol is organized into three units.

(1) The assault unit engages the enemy at the objective.

(2) The security unit secures the objective rally point, isolates the objective, and covers the patrol's withdrawal from the objective area.

(3) The support unit provides supporting fires for the assault unit attack, and covering fires, if required, for its withdrawal.

8103. Special Organization

Patrol units are further subdivided into teams, each of which performs essential, designated tasks.

a. The reconnaissance unit may be divided into reconnaissance and security teams.

b. The assault unit is usually divided into an assault team and one or more search/demolition teams.

c. If the support unit is to be divided into teams, each team must be given clear instructions as to what type support must be provided, at what location, and during what phase of the patrol.

d. The patrol leader may divide the security unit into security teams as deemed necessary.

8104. Task Organization

The patrol leader must organize the patrol in such a manner as to make sure that each individual, team, and unit is assigned a specific task or tasks. In addition, it is imperative that all the patrol members know how to perform the tasks assigned to all members of the patrol. This may not be possible in cases where trained technicians are required to perform certain tasks; however, the requirement for those technical tasks to be performed will almost certainly be the exception rather than the rule. The patrol leader must plan for maximum flexibility to take care of an emergency and ensure that the patrol's mission is not put in jeopardy with the loss of a couple of key members, a team, or an entire element.

Section II. Patrol Preparations

8201. Mission

The mission assigned to a patrol must be clear and oriented on one objective; more than one primary objective or indefinite missions invite confusion, casualties, and failure.

8202. Platoon Commander's Responsibilities

a. Provide Information. The platoon commander, having received his guidance from the company, will conduct a briefing with the squad leader who will be leading the patrol. During that briefing, the platoon commander will provide all the instructions, information, and guidance the patrol leader needs to plan and conduct his patrol. The briefing will include information such as—

- The mission of the patrol.
- General routes to be followed and/or areas to be avoided.
- Known or suspected enemy dispositions.
- Location and activities of friendly troops.
- Outposts or other security elements through which the patrol must pass.
- Terrain and weather conditions.
- Missions and routes of other patrols.
- Time patrol is to depart and return.
- Method of reporting information while on patrol (radio or messenger), place where messengers are to be sent, and where the patrol leader is to report upon completion of the patrol.
- Challenge and password to be used during the time the patrol is on its mission.
- Any special instructions such as essential elements of information sought by higher headquarters.

b. Provide Required Personnel and Equipment. The platoon commander will also inform the patrol leader of nonorganic personnel and equipment available to him, such as—

- Machine gun and/or tank-killer teams.
- Mortar, artillery forward observers and/or TACP personnel.
- Radio operator(s).
- Demolition men.
- Corpsmen.
- Other personnel (snipers, translators, military working dogs with their handlers, etc.).

c. Provide Other Support. The platoon commander will ensure that the patrol is provided with the necessary rations, water, ammunition, radios and batteries, maps, and any other items the patrol (including attachments) will require to complete the mission.

d. Review the Patrol Leader's Plan. Once the patrol leader has completed his plan, he will then brief the platoon commander on its contents. This discussion between the two should satisfy the platoon commander that the patrol leader has completely understood the mission and desired results of the patrol as it was briefed to him, and that the patrol order provides a workable means of accomplishing the desired goals. Also at this time the platoon commander will advise the patrol leader if he desires to inspect the patrol prior to its departure.

e. Debrief the Patrol. Upon the patrol's return, the patrol leader and other patrol members are debriefed by the platoon commander, company commander, or battalion S-2 officer. The debriefing should be conducted as soon as possible after the patrol's return, while information is still fresh in the minds of patrol members.

f. Control of the Patrol. It must be remembered that movement of troops outside friendly lines while in the defense, or away from the main body while in the offense or on a march, is conducted only with the approval of the company commander. The company commander controls the parol through the platoon commander.

8203. Patrol Leader's Preparation

During planning, the patrol leader uses the patrol steps, a series of mental and physical processes to ensure that all required events are planned for and all patrol members know their duties. The patrol steps incorporate the troop leading procedures discussed in appendix C, but are addressed in greater detail. The normal sequence for the patrol steps is listed below, but the sequence may vary depending on availability of personnel, times at which a reconnaissance can be conducted, and the extent of coordination already made by the platoon commander or company commander.

- Study the mission.
- Plan use of time.
- Study the terrain and situation.
- Organize the patrol.
- Select men, weapons, and equipment.
- Issue the warning order.
- Coordinate (continuous throughout).
- Make a reconnaissance.
- Complete detailed plans.
- Issue the patrol order.
- Supervise (continuous), inspect, rehearse, and reinspect.
- Execute the mission.

8204. Study the Mission

The patrol leader carefully studies the mission and all other information provided him by the platoon commander, making notes as he does so. In so doing, he identifies other significant tasks (implied missions), which must be accomplished, in order for the patrol to accomplish its primary mission. These implied missions are further identified as missions for the patrol's elements and teams which may require special preparation, planning, personnel, and/or equipment.

8205. Plan Use of Time

In order for the patrol leader to properly use the time allotted to him from receipt of the platoon commander's order until departure from friendly lines, he prepares a schedule which includes every event which must be done prior to departing friendly lines. In preparing the schedule, the patrol leader works backwards from the time of departure of friendly lines to the present. Some of the key events he knows must be done are rehearsals (day and night), issuing the patrol order to the patrol, inspecting the patrol, and, if the situation permits, making a reconnaissance of the patrol routes and the objective. An example of a patrol leader's schedule could be as follows:

0230-	-	Return to friendly lines/area.
2400-0230	-	Movement to friendly lines/area.
2330-2400	-	Accomplish mission and reorganize.
2230-2330	-	Recon objective area.
2000-2230	-	Movement to objective rally point.
2000-	-	Depart friendly lines/area.
1945-2000	-	Movement to departure point.
1930-1945	-	Final inspection.
1845-1930	-	Night rehearsals.
1800-1845	-	Day rehearsals.
1745-1800	-	Inspection.
1700-1745	-	Chow.
1630-1700	-	Issue patrol order.
1530-1630	-	Complete detailed plans.
1430-1530	-	Make reconnaissance.
1330-1415	-	Preliminary planning.
	-	Coordinate.
	-	Issue warning order.
	-	Select personnel, weapons, equipment.
	-	Organize the patrol.
	-	Study terrain and situation.
	-	Plan use of time.
	-	Study the mission.
-1330	-	Receive platoon commander's order.

8206. Study the Terrain and Situation

The patrol leader makes a thorough map study of the terrain over which the patrol will operate. The terrain in the vicinity of the objective influences the patrol organization, the manner in which the patrol leader will conduct his reconnaissance, and the disposition of his patrol at the objective. The terrain also influences the size, organization, and equipment of the patrol.

The patrol leader studies the friendly and enemy situations and determines the effect that troop dispositions, strengths, and capabilities may have on his mission. These factors influence the routes he will use, organization of the patrol, and the weapons and equipment to be taken.

8207. Organize the Patrol

Organization consists of determining the units and teams required to accomplish essential tasks. Organization is a two-step process — general organization and special organization.

8208. Select Men, Weapons, and Equipment

a. Patrol Members. Normally, the patrol leader is limited to selecting patrol members from within his own squad, with the addition of those personnel/teams made available by the platoon commander. As a rule of thumb, if the patrol leader is a squad leader, everyone in the squad will participate in the patrol. If the entire squad will not be going on the patrol, the patrol leader maintains fire team integrity whenever practicable.

b. Weapons. Patrol members are armed with organic weapons. The patrol leader should request the support of specific weapons teams when required for tasks beyond the capability of the squad.

c. Equipment. There are five general purposes or areas for which the patrol leader chooses equipment.

(1) **In the Objective Area.** This is the equipment with which the patrol accomplishes its mission. It includes such items as ammunition (number of rounds per man), demolitions (type and amount), binoculars, night vision devices, listening devices, trip flares, and flashlights.

(2) En Route. This is the equipment which enables or assists the patrol in reaching its objective. It includes such items as maps, compasses, binoculars, wire cutters, ropes, flashlights, ammunition (number of rounds per man), and boats.

(3) Control. This equipment is used in assisting the patrol leader and unit leaders in controlling the patrol while moving and during actions at the objective area. It includes such items as whistles, pyrotechnics, radios, flashlights, and luminous tape.

(4) Routine Equipment. This is the equipment carried by all patrol members. It includes the uniform to be worn and individual clothing and equipment to be carried.

(5) Water and Food. The patrol leader will specify the number of full canteens to be carried by all patrol members. Rations are rarely issued for use on patrol; however, on long patrols rations may be specified.

8209. Issue the Warning Order

Thus far the patrol leader has been going through the mental processes necessary for him to arrive at some initial conclusions about the patrol. He has determined how he will organize the patrol, what attachments he will need, determined a time schedule, made a thorough map reconnaissance, and has identified some implied missions that will have to be accomplished if the patrol's mission is to succeed. Now, he must alert his patrol members so they can begin their preparations.

Ideally, the patrol leader issues the warning order to all patrol members, including attachments. Often that is not possible. In that case the patrol leader must ensure that all unit leaders are present to receive the warning order.

The minimum items of information for inclusion in the warning order and a convenient format for arranging the information is shown in figure 8-1.

```
                    PATROL WARNING ORDER

1. A brief statement of the situation.
2. Mission of the patrol.
3. General instructions:
   a. General and special organization.
   b. Uniform and equipment common to all.
   c. Weapons, ammunition, and equipment.
   d. Chain of command.
   e. A time schedule for the patrol's guidance.
   f. Time, place, uniform, and equipment for receiving the patrol order.
   g. Times and places for inspections and rehearsals.
4. Specific instructions:
   a. To subordinate leaders.
   b. To special purpose teams or key individuals.
```

Figure 8-1. Patrol Leader's Warning Order.

a. Situation. A brief statement of the enemy and friendly situation.

b. Mission. The patrol leader reads the mission exactly as he received it.

c. General Instructions

(1) **General and Special Organization.** General tasks are assigned to units and teams. Specific details of tasks are given in the patrol leader's order.

(2) **Uniform and Equipment Common to All.** The patrol leader specifies camouflage measures to be taken and the identification to be carried.

(3) **Weapons, Ammunition, and Equipment.** These items are assigned to units and teams. Subordinate leaders make further assignments to teams and individuals.

(4) **Chain of Command.** A chain of command is established when the patrol includes personnel from outside the squad.

(5) **Time Schedule.** The patrol leader addresses all events from the present until the patrol departs. He also designates the place and uniform for receiving the patrol order, conducting inspections, and rehearsals.

d. Specific Instructions

(1) To Subordinate Leaders. The patrol leader gives out all information concerning the drawing of ammunition, equipment, ordnance, water, and rations; identifies the personnel he wants to accompany him on his reconnaissance; and gives guidance on any special preparation he believes will be necessary during the conduct of the mission, such as practicing stream crossings.

(2) To Special Purpose Teams or Key Individuals. The patrol leader should address requirements of designated personnel or teams, such as having point men, pacers, and navigators make a thorough map study and check their equipment.

8210. Coordinate (Continuous Throughout)

The patrol leader begins his coordination from the time he receives the order. He is primarily concerned with:

a. Movement in Friendly Areas. The patrol leader finds out the location of other friendly units or patrols so his patrol will not be restricted or endangered in its movement; he plans his routes and fires accordingly.

b. Departure and Reentry of Friendly Lines/Areas. The patrol leader checks with the small-unit leaders occupying the areas through which the patrol will depart and return. He ensures that they know about his patrol, times of its departure and return, and whether or not guides from their units will be required to lead the patrol through any friendly obstacles, such as mines or wire.

c. Fire Support. During his briefing with the platoon commander, the patrol leader finds out what fire support is available to him during the patrol. He then finds out what artillery and mortar targets have already been plotted along the routes to and from the objective area and within the objective area itself. Next, he plans for additional fires (if necessary) along the patrol's route to the objective area, at the objective area, and to cover his withdrawal from the objective area back to his unit's position.

d. Logistic Support. The patrol leader must arrange for the delivery or pick up of ammunition, special equipment, demolitions, water, binoculars, etc. He must also inquire as to the use of helicopters for casualty evacuation during the patrol's movement and at the objective area.

e. Information Checklist. The patrol leader should try to find out as much information about the enemy as possible. Specifically, he should determine the enemy's pattern of operation—has he been conducting patrols, what type weapons does he have, what is his strength and disposition, does he use mines and boobytraps, etc. If the patrol leader has access to him, the battalion S-2 may be able to provide some valuable information about the enemy, including some information gleaned from prior patrol reports, aerial photographs, reconnaissance operations, etc.

8211. Make Reconnaissance

Whenever possible, the patrol leader makes a physical reconnaissance of the routes he wants to follow and of the objective area. Often, because of the enemy situation, he is not able to do so, and must rely instead on his map reconaissance and information he is able to gather from other sources.

8212. Complete Detailed Plans

The patrol leader is now ready to plan his patrol in detail. Through his discussions with the platoon commander and his coordination efforts, the patrol leader has already determined the situation and the mission (paragraphs 1 and 2 of the five-paragraph order). The remainder of his planning deals with how the patrol is to be executed and the tasks assigned to each element/team (paragraph 3); administrative and logistic matters (paragraph 4); and command and signal (paragraph 5).

a. Specific Duties of Elements, Teams, and Individuals. The warning order assigned tasks to elements, teams, and key individuals. The patrol leader now assigns specific duties to each.

b. Route and Alternate Route. The patrol leader selects the patrol routes based on his map study, aerial photographs, his own reconnaissance, and/or consultation with others who have been over the terrain. He chooses a route which affords concealment from enemy observation, where little or no enemy opposition is expected, and yet presents a minimum of obstacles to the patrol. For a night patrol, the route should normally be planned to avoid thick undergrowth, dense woods, and ravines. Whenever practicable, the patrol leader should plan the return via a different route. Patrol routes are pointed out to the patrol members by—

- Indicating the routes on a map or overlay.
- Designating objectives and checkpoints.

c. Conduct of the Patrol. The patrol leader's plan must address all the following:

- Patrol formation and order of movement.
- Departure from and reentry to friendly lines or areas.
- Rally points and actions at rally points.
- Final preparation position and actions at that position.
- Objective rally point and actions to be taken at that point.
- Actions at danger areas.
- Actions on enemy contact.
- Actions at the objective.

d. Arms and Ammunition. The patrol leader checks to see if the arms and ammunition specified in the warning order have been obtained.

e. Uniform and Equipment. The patrol leader checks to see if all required equipment was available and was drawn.

f. Wounded and Prisoners. The procedures for handling wounded may vary, depending on the seriousness of the wound and if it occurs en route to the objective, at the objective, or on the return to friendly areas. A patrol may continue on to the objective carrying its casualties, may send them back with a detail of men, the entire patrol may return with the casualties, or the patrol may have to call on its parent unit for assistance. Personnel who become casualties at the objective area or on the return to friendly areas will normally be transported by whatever means are available—carried by the patrol, in vehicles, or by helicopter. Prisoners normally travel with the patrol, guarded by personnel so designated in the patrol order.

g. Signals. The patrol leader's plan addresses the type of signals to be used during the patrol—arm-and-hand signals, radio, pyrotechnic, audio. Normally, he should designate a primary and alternate signal for each event requiring signals.

h. Communication With Higher Headquarters. The patrol leader includes all essential details of communication—call signs, frequencies, reporting times (usually upon reaching checkpoints), code words, and security requirements.

i. Challenge and Password. In addition to the parent unit's challenge and password, the patrol leader designates a challenge and password to be used within the patrol, outside of friendly lines/areas.

j. Location of Leaders. The patrol leader plans his location where he can best control the patrol, usually in the forward one-third of the formation. The assistant patrol leader is placed where he can best assist in control during movement, usually near the rear of the formation. At the objective, the assistant patrol leader positions himself so he can readily take command if the patrol leader becomes a casualty.

8213. Issue Patrol Order

When the patrol leader has completed his plan (see fig. 8-2), he assembles the members of the patrol and issues the order. He issues the order in a clear, concise manner, following the standard five-paragraph order format. He should—

- Ensure all patrol members are present.
- Receive a status report from his unit/team leaders on the prepatory tasks assigned to them when he issued the warning order.
- Precede the issuance of the order with an orientation.
- Build a terrain model using dirt, rocks, twigs, etc., to help explain the concept of operations for movement to the objective area, actions at the objective area, and the return to friendly lines/area.
- Issue the entire order before taking questions.
- Conclude the question/answer session with a time check and announce the time of the next event. ("It is now 1700. Everyone get some chow and I'll inspect the patrol, in movement formation, at 1745, in that clump of pines near the company CP".)

1. SITUATION

    a. <u>Environment</u>. Weather, terrain, visibility; local population situation and behavior as they impact on the patrol and enemy forces.

    b. <u>Enemy Forces</u>. Identification, location, activity, strength.

    c. <u>Friendly Forces</u>. Mission of next higher unit, location and planned actions of adjacent units, mission and routes of other patrols, availability of supporting fires and other support.

    d. <u>Attachments and Detachments</u>. Time and units affected.

2. MISSION

    A clear concise statement of the task which the patrol must accomplish.

3. EXECUTION

    a. <u>Concept of Operations</u>. The concept tells the where, how, and who and lays out the patrol leader's general scheme for accomplishing the mission. It outlines the following:

    (1) Task organization of the patrol.

    (2) Movement to the objective area, to include navigation method.

    (3) Actions in the objective area.

    (4) The return movement, to include navigation method.

    (5) Use of supporting forces (including illumination, if required).

    b. <u>Tasks</u>. Missions are assigned to elements, teams, and individuals, as required.

    c. <u>Coordinating Instructions</u>. This paragraph contains instructions common to two or more elements, coordinating details, and control measures applicable to the patrol as a whole. At a minimum, it will include:

    (1) Time of assembly in the assembly area.

    (2) Time of inspections and rehearsals (if not already conducted).

    (3) Time of departure and estimated time of return.

    (4) Location of departure and reentry of friendly lines and the actions associated with departure and reentry.

    (5) Details on the primary and alternate routes to and from the objective area.

    (6) Details on formations and order of movement.

    (7) Rally points and actions at rally points.

    (8) Final preparation position and actions at this position.

    (9) Objective rally point and actions at this point.

    (10) Actions at danger areas.

    (11) Actions in the event of enemy contact.

    (12) Details on actions in the objective area not covered elsewhere.

    (13) Estimated time of patrol debriefing upon return.

> 4. ADMINISTRATION AND LOGISTICS
>
>   a. Changes/additions to uniform, equipment, and prescribed loads from that given in the warning order.
>
>   b. Instructions for handling wounded and prisoners.
>
> 5. COMMAND AND SIGNAL
>
>   a. <u>Command Relationships</u>. Chain of command and succession to command.
>
>   b. <u>Signal</u>. Challenge and password, arm/hand and special signals, and radio frequencies and call signs.
>
>   c. <u>Command Posts</u>. Position of patrol leader and assistant patrol leader within the patrol organization during the approach and return and at the objective.

Figure 8-2. Patrol Order.

8214. Supervise (Continuous), Inspect, Rehearse, and Reinspect

Inspections and rehearsals are vital to proper preparation. They are conducted even when the patrol leader and patrol members are experienced in patrolling.

a. Inspections determine the state of readiness of the men, both mental and physical.

(1) The patrol leader inspects the patrol just before conducting rehearsals. He looks for —

- Prescribed uniform, weapons, ammunition, ordnance, and equipment.
- Camouflage.
- Identification tags and Geneva Convention cards.
- Unnecessary equipment and personal items.

(2) The patrol leader questions each member of the patrol to ensure he knows —

- The mission, routes, and fire support plan.
- His assignment and during what part of the patrol he performs it.
- What other members of the patrol are to do at certain times during the patrol.
- Challenges and passwords, call signs, frequencies, code words, reporting times, and other pertinent details.

(3) If there is any time between the final rehearsal and the time to depart, the patrol conducts another inspection.

b. Rehearsals ensure the operational proficiency of the patrol. Plans are checked and any necessary changes are made. The patrol leader verifies the suitability of the equipment. It is through rehearsals that patrol members become thoroughly familiar with the actions they are to take during the patrol.

(1) If the patrol is to operate at night, conduct both day and night rehearsals. They should be conducted on terrain similar to that on which the patrol will operate. All actions should be rehearsed. If time is limited, only the most critical phases should be rehearsed. Action at the objective area is the most critical phase and should always be reheased.

(2) The patrol leader should talk the patrol through each phase, describing the actions and having each man perform his duties. He should then walk the patrol through all phases of the patrol, using only signals and commands which will be used during the actual patrol.

(3) When rehearsals are completed and the patrol leader is satisfied with the members' performance, he makes any final adjustments to his plan or patrol organization. He then issues final instructions to his unit/team leaders, noting any changes he has made. While his subordinate leaders are giving the final instructions to their men, the patrol leader informs the platoon leader that the patrol is ready to depart.

Section III. Conduct of Patrols

8301. Formation and Order of Movement

In organizing the patrol for movement, the patrol leader determines the formation(s) in which the patrol will move to the objective area. He also determines the location of units, teams, and individuals in the formation. As far as practicable, the patrol leader organizes the movement so as to maintain unit and team integrity.

The standard squad and fire team formations are adaptable to any patrol. The patrol leader may change from one formation to another depending on the situation. Other considerations impacting on the patrol formations are:

- Probability of contact with the enemy.
- Terrain, weather, vegetation, and visibility.
- Time allotted for the patrol to accomplish its mission and return to friendly lines/areas.

8302. Departure and Reentry to Friendly Lines/Areas

a. During his preparation phase, the patrol leader makes contact with the leaders of the units occupying the areas through which the patrol must depart and reenter friendly lines/areas. In some instances coordination is done at higher levels and the patrol leader is simply told where he is to depart and return. In either case, the patrol leader should try to have a face-to-face session with the unit leaders to reduce the possibility of mistakes occurring during the passage of lines.

b. The patrol leader also picks out an assembly area for his patrol to move into while he coordinates final preparations with the unit through which the patrol will be moving. If possible, the patrol's route from its assembly area through friendly positions should offer concealment from enemy view.

c. Once the patrol leader has received permission from the platoon commander to move out, he leads his patrol directly from its rehearsal area to the assembly area. Leaving his assistant patrol leader in charge, the patrol leader, usually accompanied by a radio operator or messenger, moves to the unit occupying the positions through which the patrol will depart. There he meets with the personnel with whom he has previously coordinated (company commander, platoon commander, or squad leader), and makes final arrangements for departure. If guides are required to lead the patrol through the unit's position and/or its local security area, the patrol leader picks them up at this time.

d. Once the departure has been coordinated, the patrol leader advances the patrol to his position. He may call the patrol forward by radio, send his messenger back to the patrol's assembly area to lead it forward, or go back himself to lead the patrol forward.

e. With the patrol assembled in its formation for movement, the patrol leader or the guide leads the patrol through the friendly positions. At the last friendly position, the patrol leader dismisses the guide to return to his own unit. He also takes this time to give any last, very brief instructions to the patrol's point or navigator if he considers it necessary. As the patrol moves past him, the patrol leader takes his position in the formation for the patrol. Usually the patrol leader transmits to his platoon commander or company commander that the patrol has departed friendly lines/area.

f. Also, during his planning, the patrol leader effects coordination with the leaders of the unit through which the patrol will return. He provides them information about the size of the patrol, general route, and expected time of return. The manner of challenge and recognition of the returning patrol is coordinated in great detail.

g. Upon returning to friendly forward local security area or front lines, the patrol leader leaves the patrol in a covered position while he and a radio operator or messenger move forward to make contact with the friendly unit in the manner previously agreed upon. Once he has made contact, the patrol leader calls the patrol forward, sends his messenger back, or returns himself to lead the patrol forward to the passage point. It is imperative that the patrol leader check each man personally as he reenters.

8303. Exercise of Control

a. The patrol leader positions himself where he can best control the patrol as a whole. The assistant patrol leader moves at or near the rear and prevents straggling. Other unit and team leaders move with their units. All patrol members must stay alert and pass on signals and instructions. A signal to halt may be given by any patrol member, but only the patrol leader may give the signal to resume.

b. Arm-and-hand signals are the primary means of communication within a patrol and should be used exclusively when near the enemy. All members must know standard infantry signals as well as any special signals and be alert to receive and pass them to other members.

c. The patrol leader should speak just loudly enough to be heard. At night, or when close to the enemy, he halts the patrol and has subordinate leaders come forward. He gives the information to them and they then pass it on to their subordinates by moving quietly from man to man.

d. Radios provide a means of positive control within a large patrol, but should only be used when arm-and-hand signals or face-to-face communication is impractical.

e. The patrol leader may designate other sound signals if he can be sure they will serve their intended purpose. Sound signals must be natural sounds that are easily understood. If used, they must be planned for and rehearsed, keeping in mind that fewer signals are better.

f. Night vision devices, if available, are excellent aids in exercising control of the patrol. Also, small strips of luminous tape on the back of the cap or collar of patrol members can aid in keeping visual contact with the man in front.

g. An important aspect of control is personnel accountability. Personnel must be accounted for after crossing danger areas, after halts, and after enemy contact.

   (1) When moving in a column, the patrol leader turns to the man behind him and says, "Send up the count". This is passed back to the last man who starts the count. The last man sends up the count by tapping the man in front of him and saying "one" in a low voice. This man taps the man in front of him and says, "two". This continues until the count reaches the patrol leader. The men behind the patrol leader, plus the patrol leader, and the men he knows to be ahead of him, should equal the total of the patrol.

(2) After enemy contact or after dispersal and reassembly at a rally point, the patrol leader or senior man obtains a count by the quickest method available. Time and the situation permitting, he should go from man to man himself. This also gives him the opportunity to check on the condition of the men.

(3) The patrol leader may give guidance in his patrol order to send up the count automatically at various times or after certain events/occurrences during the patrol.

8304. Navigation

One or more men in the patrol are assigned as navigators. Their function is to assist the patrol leader in maintaining direction by use of the compass. The patrol leader also assigns men as pacers to keep track of the distance from point to point. He should assign at least two pacers and use the average of their counts for an approximation of the distance traveled. The pacers are separated so they will not influence each other's count.

The route is divided into legs, with each leg starting at a recognizable point on the ground. The pacers begin their counts from zero at the beginning of each leg. It may be convenient to use the distance from checkpoint to checkpoint as a leg. Periodically, the patrol will call for a pace count. The count from both pacers is sent up in the same manner as the personnel count. All patrol members must understand that the count of both pacers must be sent forward, and the counts will be different.

8305. Security

The patrol leader organizes the formation to provide security while on the move, during halts, at danger areas, and upon reaching checkpoints and rally points. Scouts are always employed to the front and rear of the formation. Consistent with visibility, the terrain, and vegetation, scouts are employed to the flanks. When employment of flank scouts is impractical, they move with the patrol formation, but maintain observation to their assigned flanks. Depending on the size of the patrol, scouting units may consist of one or two men, or a fire team. Regardless of the size of the scouting units and where they are employed, they must maintain visual contact with the patrol leader at all times.

a. Day Patrols

(1) The patrol is dispersed consistent with control, visibility, cover, concealment, and the enemy situation.

(2) Scouts are employed to the front and rear, and to the flanks, if practical.

(3) The patrol moves so as not to silhouette itself when moving along high ground.

(4) Movement is along available covered and concealed routes, and exposed areas are avoided whenever possible.

(5) The patrol avoids known or suspected enemy locations and built-up areas.

(6) The patrol maintains an even pace. Sudden movements attract attention.

b. Night Patrols

(1) Patrol members stay closer together.

(2) Silent movement is emphasized.

(3) Speed of movement is slowed to reduce the possibility of men becoming separated from the patrol.

c. Halts. The patrol leader halts the patrol occasionally to observe and listen for enemy activity. During these security halts, on signal from the patrol leader, every man freezes in place, remains quiet, observes, and listens. Security halts are called upon reaching danger areas and periodically along the route. When the patrol gives the signal to resume movement, the signal is passed rearward. The last man, upon receiving the signal, gives a "thumbs-up" which is passed from man to man forward to the patrol leader. Once the patrol leader gets the "thumbs-up" from the men to his rear, he signals forward to resume movement.

8306. Movement Control Measures

a. Checkpoints. A checkpoint is a predetermined point on the ground used as a means of controlling movement. During his map study or physical reconnaissance, the patrol leader decides the number and locations of checkpoints to be plotted along the patrol route. These are coordinated with his parent unit before the patrol leaves. Checkpoints are assigned numbers, not in sequential order. Normally, the patrol leader will call in upon reaching checkpoints so that the parent unit will be able to follow the progress of the patrol toward its objective and on its return to friendly lines/area.

b. Rally Points. A rally point is an easily identifiable point on the ground, designated by the patrol leader, where the patrol can reassemble/reorganize if it becomes dispersed. It should provide cover and concealment and be defensible for a short time. All rally points are considered tentative until they are reached, found to be suitable, and designated by the patrol leader. He ensures that all patrol members are notified when a rally point is so designated, either by arm-and-hand signal or by passing the word orally. He also points out identifying features which mark the limits of the rally point.

(1) **Types of Rally Points**

(a) **Initial Rally Point.** This is a point within the friendly area where the patrol can reassemble if it becomes dispersed before departing the friendly area or before reaching the first rally point designated en route. It may be the patrol assembly area. The initial rally point location must be coordinated with the commander in whose area it lies.

(b) En Route Rally Points. These are points selected along the patrol's route to the objective and from the objective back to friendly lines/area. The patrol leader selects them as the patrol passes through likely areas for which rally points are needed.

(c) Objective Rally Point. This is the rally point nearest the objective at which the patrol reassembles after the mission is accomplished. It may be located short of, to a flank, or beyond the objective. This may also be used as the final preparation point (see below).

(2) Selecting Rally Points. The patrol leader selects likely locations for tentative rally points during his map study or reconnaissance. He always selects a tentative initial rally point and tentative objective rally point, and identifies them in his patrol order. He also selects additional rally points along the patrol route, if necessary. Rally points should always be designated on both the near and far sides of danger areas, such as roads, trails, open areas, and streams.

(3) Actions at Rally Points. The patrol leader plans the actions to be taken at rally points and instructs the patrol accordingly during the patrol order. Planned actions at the initial rally point and en route rally points must provide for the continuation of the patrol as long as there is a reasonable chance to accomplish the mission. Actions to be taken at rally points should address—

- Recognition signals for assembly at rally points.
- The minimum number of men required and the maximum waiting time before the senior man at the rally point moves the rallied patrol members on to the objective.
- Instructions for patrol members who find themselves alone at a rally point.

(4) Final Preparation Position. The final preparation position is that location in the vicinity of the objective where the patrol makes the final preparations prior to approaching the objective. This position must provide the patrol concealment from enemy observation and, if possible, cover from enemy fires. It is this position from which the patrol leader's reconnaissance is made; it serves as the release point from which units and teams move into position to accomplish the mission at the objective. If the situation permits, the final preparation point can also be used as the objective rally point. Using the final preparation position as the objective rally point provides the patrol with the advantage of leaving behind clothing and equipment not required at the objective.

8307. Actions at Danger Areas

A danger area is any place where the patrol is vulnerable to enemy observation or fire (e.g., open areas, roads, trails, and obstacles such as barbed wire, minefields, streams, and lakes). Any known or suspected enemy position the patrol must pass is also a danger area. The patrol leader plans for crossing each danger area identified during his reconnaissance or map study, and includes the plans in his patrol order.

When a patrol approaches a danger area, the near side of the danger area is reconnoitered first; then the patrol leader sends scouts to reconnoiter the far side. The scouts should also reconnoiter the tentative rally point on the far side. Once the scouts report that the far side is clear of enemy, the remainder of the patrol crosses the danger area. As each individual/group crosses the danger area, others provide cover. If possible, enemy obstacles are avoided as they are usually covered by fire.

In crossing a river, the near bank is reconnoitered first; then the patrol is positioned to cover the far bank as scouts are sent across. When the scouts report that the far bank is clear of enemy, the remainder of the patrol crosses as rapidly as possible. If the crossing requires swimming, the patrol uses improvised rafts to float weapons, ammunition, and equipment across the river.

Roads and trails are crossed at or near a bend or a narrow part.

A patrol may be able to take advantage of battlefield noises to help cover its movement across danger areas near enemy positions. If supporting arms are available, the patrol leader can call for them to divert the enemy's attention while the patrol crosses the danger area or passes the enemy position.

8308. Immediate Actions Upon Contact With the Enemy

a. General. A patrol may make contact with the enemy at any time. Contact may be made through observation, a meeting engagement, or an ambush. Contact may be visual, in which the patrol sights the enemy but is not itself detected. In this case, the patrol leader can decide whether to make or avoid physical contact, based on the patrol's mission and capability to successfully engage the enemy.

(1) A reconnaissance patrol's mission prohibits physical contact, except that necessary to accomplish the mission. Its actions are defensive in nature. Physical contact, if unavoidable, is broken as quickly as possible and the patrol, if still capable, continues its mission.

(2) A combat patrol's mission is to seek or exploit opportunities for contact. Its actions are offensive in nature. When making enemy contact, the patrol's actions are swift and violent in an effort to inflict maximum damage on the enemy, followed by immediate relocation to another area, or return to friendly lines/area. Patrols can expect to make physical contact with the enemy either in a meeting engagement or an ambush.

(a) A meeting engagement is a combat action that occurs when a patrol, which is incompletely deployed for battle, engages the enemy at an unexpected time and place. It is an accidental meeting where neither the patrol nor the enemy expect contact and are not specifically prepared to deal with it.

(b) An ambush is a surprise attack by fire from concealed positions on a moving or temporarily halted force.

b. Immediate Action (IA) Drills. During a patrol, contacts are often unexpected, occur at very close ranges, and are short in duration. Enemy fire may allow leaders little or no time to evaluate situations or give orders. In these situations, IA drills provide a means for swiftly initiating positive offensive or defensive action, as appropriate. IA drills are designed to provide swift and positive small-unit reaction to visual or physical contact with the enemy. They are simple courses of action which can be initiated by using minimal signals or commands. All patrol members must be trained as a unit in conducting IA drills; even then, all drills must be rehearsed prior to going on patrol. It is not feasible to design an IA drill for every possible situation. It is better to know one drill for each of a limited number of engagements or occurrences.

c. Immediate Halt Drill. When the situation requires the immediate, in-place halt of the patrol, the immediate halt drill is used. It is used when the patrol detects the enemy but is not itself detected. The first man detecting the enemy gives the arm-and-hand signal to **FREEZE**. Every man halts in place, weapon at the ready, and remains absolutely motionless and quiet until further signals or orders are given.

d. Air Observation and/or Attack Drills. These drills are designed to reduce the danger of detection from aircraft and casualties from air attack.

(1) Air Observation. When an unidentified or known enemy aircraft which may detect the patrol is heard or seen, the appropriate IA drill is **FREEZE**.

(2) Air Attack. When an aircraft detects the patrol and makes a low level attack, the IA drill for air attack is used. The first man sighting an attacking aircraft shouts, "Aircraft, front (rear, left, right)". The patrol moves quickly into line formation, well spread out, at right angles to the aircraft's direction of travel. As each man comes on line, he hits the deck, using available cover. He positions his body at right angles to the aircraft's direction of travel, to present the shallowest target possible. Between attacks (if the aircraft returns or there is more than one attacking aircraft), patrol members seek better cover. Attacking aircraft are fired on only on command of the patrol leader.

e. Meeting Engagement Drills

(1) Hasty Ambush. This IA drill is both a defensive measure to avoid contact and an offensive one to make contact. It may often be a subsequent action after the command to freeze has been given. When the signal **HASTY AMBUSH** is given, the entire patrol moves quickly to the right or left of line of movement, as indicated by the signal, and takes up the best available concealed firing positions. The patrol leader initiates the ambush by opening fire and shouting, **FIRE**; thus ensuring the ambush is initiated even if his weapon misfires. If the patrol is detected before this, the first man aware of the detection initiates the ambush by firing and shouting.

(a) When used as a defensive measure to avoid contact, the hasty ambush is not initiated unless the patrol is detected.

(b) When used as an offensive measure, the enemy is allowed to advance until he is in the most vulnerable position before the ambush is initiated.

(c) An alternate means for initiating the ambush is to designate an individual (e.g., point or last man) to open fire when a certain portion of the enemy unit reaches or passes him.

(2) Immediate Assault. This IA drill is used, defensively, to make and quickly break undesired but unavoidable contact (including ambush); and, offensively, to decisively engage the enemy (including ambush). When used in a meeting engagement, men nearest the enemy open fire and shout, "Contact, front (rear, left, right)". The patrol moves swiftly into line formation and assaults.

(a) When used defensively, the assault is stopped if the enemy withdraws and contact is broken quickly. If the enemy stands fast, the assault continues through the enemy positions and further, until enemy contact is broken.

(b) When used offensively, the enemy is decisively engaged. Any enemy attempting to escape are pursued and killed or captured.

(3) Breaking Contact. Two methods used in breaking contact are the use of fire and maneuver and by using the clock system.

(a) To break contact by using fire and maneuver, one portion of the patrol returns the enemy fire while another portion moves by bounds away from the enemy. Each portion of the patrol covers the other by fire until the entire patrol breaks contact.

(b) In using the clock system to break contact, the patrol leader shouts a direction and a distance. Twelve o'clock is always the direction of movement of the patrol. If the patrol leader shouts, "Ten o'clock—two hundred", it means for the patrol to move in the direction of ten o'clock for two hundred meters. Patrol members try to keep their relative positions as they move so the original formation is disrupted as little as possible, since this will facilitate reorganization once the patrol has broken contact. Subordinate leaders must be alert to ensure that their unit and team members receive the correct order and move as directed.

f. Counterambush Drills. When a patrol is ambushed, the IA drill used is determined by whether the ambush is near (enemy within fifty meters of the patrol) or far (enemy beyond fifty meters of the patrol). Fifty meters is considered the limit from which the ambushed patrol can launch an assault against the enemy.

(1) In a *near ambush*, the killing zone is under very heavy, highly concentrated, close range fires. There is little time or space for men to maneuver or seek cover. The longer they remain in the killing zone, the more certain their deaths. If attacked from near ambush:

(a) Men in the killing zone immediately assault the enemy's position without waiting for any order or signal. The assault should be swift, violent, and destructive. The men fire their weapons at the maximum rate, throw hand grenades, and yell as loudly as possible—anything to kill as many enemy as they can, and confuse the enemy survivors. Once they reach the ambush position, they either continue with their assault, or break contact, as directed.

(b) Men not in the killing zone maneuver against the ambush force, firing in support of those assaulting.

(c) If the ambush force is small enough to be routed or destroyed, the patrol members should continue with their assault and supporting fire. If the force is well-disciplined and holds its ground, then the patrol members should make every effort to break contact as quickly as possible, and move to the last en route rally point to reorganize.

(2) In a *far ambush*, the killing zone is also under very heavy, highly concentrated fires, but from a greater range. The greater range precludes those caught in the killing zone from conducting an assault. The greater range does, however, permit some opportunity for the men to maneuver and seek cover. If attacked from far ambush:

(a) Men in the killing zone immediately return fire, take the best available cover, and continue firing until directed otherwise.

(b) Men not in the killing zone maneuver against the ambush force, as directed.

(c) The patrol leader either directs his unit and team leaders to fire and maneuver against the ambush force, or to break contact, depending on his rapid assessment of the situation.

(3) In each situation, the success of the counterambush drill employed is dependent on the men being well-trained in recognizing the nature of an ambush and well-rehearsed in the proper actions to take. Each man has to be confident in himself, his abilities, and those of his fellow Marines. He can't wait for someone to tell him what to do, as his leaders may become casualties. Training gives the Marine the confidence and ability to do whatever it takes to accomplish the mission.

Section IV. Reconnaissance Patrols

8401. General

The commander needs information about the enemy and the terrain he controls. He must have accurate and timely information to assist him in making tactical decisions. Reconnaissance patrols are one of the most reliable sources for this information. A reconnaissance patrol is capable of carrying the search for information into the area occupied by enemy forces, usually beyond the range of vision or ground observation, and is capable of examining objects and events at close range. Routinely, a reconnaissance patrol will not maintain communication with its parent unit.

8402. Missions

Missions for reconnaissance patrols include gaining information about the location and characteristics of friendly or hostile positions and installations, routes, stream/river crossings, obstacles, or terrain; identification of enemy units and equipment; enemy strength and disposition; movement of enemy troops or equipment; presence of mechanized units; presence of nuclear, biological, and chemical equipment or contaminated areas; and unusual enemy activity.

8403. Types of Reconnaissance

a. Area Reconnaissance. An area reconnaissance (previously referred to as a point reconnaissance) is a directed effort to obtain detailed information concerning specific terrain or enemy activity within a specific location. The objective of the reconnaissance may be to obtain timely information about a particular town, bridge, road junction, or other terrain feature or enemy activity critical to operations. Emphasis is placed on reaching the area without being detected.

b. Zone Reconnaissance. A zone reconnaissance (previously referred to as an area reconnaissance) is a directed effort to obtain detailed information concerning all routes, obstacles (to include chemical or biological contamination, terrain, and enemy forces within a particular zone defined by specific boundaries.

c. Route Reconnaissance. A route reconnaissance is a reconnaissance along specific lines of communications, such as a road, railway, or waterway, to provide information on route conditions and activities along the route.

(1) Reconnaissance of routes and axes of advance precede the movement of friendly forces. Lateral routes and terrain features that can control the use of the route must be reconnoitered.

(2) Considerations include trafficability, danger areas, critical points, vehicle weight and size limitations and locations of obstacle emplacements.

(3) The route reconnaissance is narrower in scope than the zone reconnaissance. The limits of the mission are normally described by a line of departure, a specific route, and a limit of advance.

8404. Task Organization of Reconnaissance Patrols

Generally, a rifle squad is used for reconnaissance patrols; other teams or individuals having specialized capabilities may be attached to the squad for the conduct of the patrol's mission. The patrol should be organized with one or two fire teams to actually conduct the reconnaissance mission and the remaining fire team to provide security. A small *point reconnaissance patrol* needs only one team for the assigned mission. An *area reconnaissance patrol* should use two teams to physically conduct the mission and one team for cover and security. The *patrol security team* should cover the likely avenues of approach into the objective, protect the units conducting the reconnaissance, and cover the objective rally point.

8405. Equipment

Patrol members are armed and equipped as necessary to accomplish the mission. The automatic rifle in each fire team provides a degree of sustained firepower in case of enemy contact.

8406. Actions at the Objective

A reconnaissance patrol tries to conduct its reconnaissance without being discovered. Stealth and patience are emphasized. The patrol fights only to accomplish its mission or protect itself. In some situations, the patrol leader can locate enemy positions by having some of his men fire to draw the enemy's fire. It is not used if there is any other way to accomplish the mission, and is used only when authorized.

a. Area Reconnaissance. The patrol leader halts and conceals the patrol near the objective in the final preparation position. The patrol leader then conducts his leader's reconnaissance to pinpoint the objective and confirm his plan for positioning the security teams and employing units assigned the reconnaissance mission. He returns to the patrol and positions the security. They are placed to provide early warning of enemy approach and secure the objective rallyng point. The reconnaissance unit(s) then reconnoiters the objective. The reconnaissance unit may move to several positions, perhaps making a circle around the objective, in order to conduct thorough reconnaissance. When the reconnaissance is completed, the patrol leader assembles the patrol and tells everyone what he has observed and heard. Other patrol members contribute anything they may have observed. The patrol then returns to the friendly area and the patrol leader makes a full report.

b. Zone Reconnaissance. The patrol leader halts the patrol at the final preparation position, conducts his leader's reconnaissance, and confirms his plan. He positions the security team and sends out the reconnaissance teams. When the entire patrol is used to reconnoiter the area, it provides its own security. After completing the reconnaissance, each reconnaissance team moves to the objective rallying point and reports to the patrol leader. The patrol then returns to the friendly area and makes a full report.

c. Route Reconnaissance. The patrol leader halts and conceals the patrol near the objective in the final preparation position. The patrol leader then conducts his leader's reconnaissance to pinpoint the objective and confirm his plan for positioning the security teams and employing units assigned the reconnaissance mission. He returns to the patrol and positions the security. They are placed to provide early warning of enemy approach and secure the objective rallyng point. The reconnaissance unit(s) then reconnoiters the objective (route). The reconnaissance unit may move to several positions, along or adjacent to the specific route, in order to conduct a thorough reconnaissance. After completing the reconnaissance, each reconnaissance team moves to the objective rallying point and reports to the patrol leader. The patrol then returns to the friendly area and makes a full report.

Section V. Combat Patrols

8501. General

Combat patrols are assigned missions which usually require them to actively engage the enemy. As a secondary mission, they collect and report information about the enemy and terrain. Combat patrols are employed in both offensive and defensive operations. Combat patrols can inflict damage on the enemy, establish or maintain contact with friendly or enemy forces, deny the enemy access to key terrain, probe enemy positions, and protect against surprise and ambush.

8502. Types of Combat Patrols and Their Missions

a. Raid Patrols. Raid patrols destroy or capture enemy personnel or equipment, destroy installations, or free friendly personnel who have been captured by the enemy.

b. Contact Patrols. Contact patrols establish and/or maintain contact with friendly or enemy forces.

c. Economy of Force Patrols. Economy of force patrols perform limited objective missions such as seizing and holding key terrain to allow maximum forces to be used elsewhere.

d. Ambush Patrols. Ambush patrols conduct ambushes of enemy patrols, carrying parties, foot columns, and convoys.

e. Security Patrols. Security patrols detect infiltration by the enemy, kill or capture infiltrators, and protect against surprise or ambush.

8503. Task Organization of Combat Patrols

As with a reconnaissance patrol, the combat patrol is task-organized to perform the specific mission assigned. FMFM 6-7, *Scouting and Patrolling for Infantry Units*, contains detailed discussion on combat patrols. As with other combat missions, the size and organization of a combat patrol depends on the mission, enemy situation, troops available to conduct the patrol, and terrain. The commander ordering the patrol must evaluate all these elements in arriving at his decision regarding the type of patrol and what unit will conduct it.

8504. Equipment

Combat patrols are armed and equipped as necessary for accomplishing the mission. In addition to binoculars, wire cutters, maps, compasses, night vision devices, and other equipment common to all patrols, the combat patrol is generally armed with much greater firepower than is the rifle squad. As success of the mission may depend on the patrol's ability to call for supporting arms fire, radio communications plays a much more important role than in the reconnaissance patrol. The patrol must be able to communicate with higher headquarters, and radio communications among units/teams should be provided.

8505. Contact Patrols

a. General. Contact patrols establish and/or maintain contact to the front, rear, or flanks by:

(1) Contacting friendly forces at designated contact points. (A contact point is an easily identifiable point on the ground where two or more units are required to make contact. The order establishing the patrol should state what contact is required; e.g., physical, visual, radio.)

(2) Establishing contact with a friendly or enemy force when the definite location of the force is unknown.

(3) Maintaining contact with friendly or enemy forces, without becoming decisively engaged with the enemy.

b. Organization and Equipment. A contact patrol's organization and equipment is dependent on the known enemy situation and anticipated enemy contact.

(1) A patrol operating between adjacent friendly units, making contact at designated points, is usually small and relatively lightly armed.

(2) A patrol sent out to establish contact with an enemy force is organized, armed, and equipped to overcome resistance offered by light screening forces, in order to gain contact with the main enemy force. It is not organized and equipped to engage the main enemy force in combat.

(3) Reliable radio communications over the entire distance covered by the patrol must be provided.

c. Actions at the Objective. The patrol leader selects a series of objectives. If his mission is to gain or maintain contact with friendly forces, these objectives may double as contact points. If the mission is to gain or maintain contact with the enemy, the objectives may be terrain features, an enemy screening force, or the main enemy force. The patrol leader initially selects probable objectives during his preparation time. Once on patrol, he will select objectives while on the move, depending on what the enemy does. His mission will dictate his choice of objectives. If the mission is to keep the enemy under surveillance, his objectives will be terrain features from which he can do so. If his mission is to maintain pressure on the enemy, his objective may be the enemy screening force, and he will continually deploy his men to conduct a series of attacks against that force, reorganizing the patrol after each attack. If the enemy reacts strongly, the patrol leader should withdraw his force and seek another time or place from which to again put pressure on the enemy force. Above all, the patrol leader must take care not to become decisively engaged with the enemy.

8506. Ambush Patrols

An ambush is a surprise attack from a concealed position upon a moving or temporarily halted target. It is one of the oldest and most effective types of military action. The ambush may include an assault to close with and decisively engage the enemy, or the attack may be by fire only.

a. Purpose of Ambushes. Ambushes are executed for the general purpose of reducing the enemy's overall combat effectiveness and for the specific purpose of destruction of his units. The cumulative effect of many small ambushes on enemy units lowers the morale of enemy troops and, in general, is a harassment to the enemy force as a whole.

(1) The primary purpose of ambushes is to kill enemy troops.

(2) Harassment is a secondary purpose for conducting ambushes. Frequent ambushes force the enemy to divert men from other missions to guard convoys, troop movements, etc. When enemy patrols are ambushed, they fail to accomplish their mission, and the enemy is deprived of valuable information. Successful ambushes cause the enemy to be less aggressive and more defensive minded. The enemy troops become apprehensive and overly cautious, they become reluctant to conduct night operations and patrols, and are more subject to confusion and panic if they are ambushed. In general, they become less effective fighters.

b. Classification of Ambushes. Ambushes are classified as deliberate ambushes or ambushes of opportunity. A deliberate ambush is one that is planned against a specific target. Detailed information is required about the target — the nature of the target, its size and organization, armament and equipment, the route it will follow, rate of movement, and times it will reach or pass certain identifiable points along its route. A deliberate ambush is normally conducted by a reinforced platoon. A deliberate ambush may be planned for such targets as—

- Logistic columns, either rail or motorized.

- Troop movements, either rail, motorized, or on foot.

- Enemy patrols which establish patterns by frequently using the same routes or habitually depart and reenter their own areas at the same point.

- Any other force, when sufficient prior information is known.

When detailed information necessary to conduct a deliberate ambush is not available, but there are routes that the enemy will likely be using, day or night, for troop movements, patrols, or convoys, the commander may order an ambush patrol, or a number of ambushes, to cover those routes. These are the ambush patrols most often conducted, and the reinforced rifle squad is well-suited to perform them. Generally, the patrol will be given a mission to organize an ambush along a road or trail near a specific point or area, and execute the ambush against the first profitable target that appears.

c. Types of Ambushes. There are two types of ambushes — point and area.

(1) The *point ambush* is one where forces are deployed to attack along a single killing zone.

(2) The *area ambush* is one where forces are deployed as multiple related point ambushes. A squad may conduct a point ambush as part of a larger area ambush; it cannot, however, conduct an area ambush on its own. This holds true even when the squad is reinforced with crew-served weapons.

d. Factors for a Successful Ambush

(1) Routes. A route is planned which will allow the patrol to enter the ambush site from a direction that will not take it through the planned killing zone. If the killing zone must be entered or crossed (e.g., to place mines), patrol members must take care not to disturb the natural appearance of the area. Nothing should be done that could alert the enemy and compromise the ambush. A covered and concealed withdrawal route is also planned. As with all patrols, alternate routes are planned, as necessary.

(2) Ambush Site. In conjunction with the map or aerial photo study conducted by the patrol leader during his preparation phase, a physical reconnaissance of the site must be made. The patrol leader must remember that a suitable ambush site for his patrol is also a suitable site for one by the enemy. Additionally, it must be looked at first-hand to ensure that what looked good on the map is actually a good site for an ambush. An ambush site must provide for—

(a) Favorable fields of fire.

(b) Canalization of the target into the killing zone. An ideal killing zone restricts the enemy on all sides, confining him to an area where he can quickly and completely be destroyed. A killing zone flanked by natural obstacles such as cliffs, streams, embankments, or steep grades, supplemented by other obstacles such as mines or barbed wire, limit an enemy's opportunity to escape or to employ his own counter-ambush drills against the ambush.

(3) Positions. The security teams move into position first, to prevent the patrol from being surprised as it moves into the ambush site. Automatic weapons and machine guns are positioned so that each can fire along the entire killing zone, or their sectors of fire must overlap so that the killing zone is covered completely. Riflemen and grenadiers are positioned to cover any dead space left by the automatic weapons and machine guns. The patrol leader then selects a position from which he can best initiate the ambush. He then gives instructions as to clearing fields of fire, preparing positions, and camouflage, and sets a time by which all preparations are to be completed.

(4) Local Security. The security unit, normally employed in two teams, does not usually participate in the initial attack, but protects the flanks and rear, giving early warning of enemy approach, and covering the patrol's withdrawal at the conclusion of the ambush.

(5) Surprise. Surprise distinguishes the ambush from other forms of attack. It is surprise that allows the ambush patrol to seize and retain control of the situation. If complete surprise cannot be achieved, it must be so nearly complete that the target is not aware of the ambush until too late for effective reaction.

(6) Coordinated Fires. All weapons, mines, and demolitions must be positioned, and all fires, including those of available artillery and mortars, must be coordinated to achieve:

- Isolation of the killing zone to prevent escape or reinforcement.
- Surprise delivery of a large volume of highly concentrated fires into the killing zone.

(7) Control. Close control must be maintained during movement to, occupation of, and withdrawal from the ambush site. This is best achieved through rehearsals and maintenance of good communications. The patrol members must control themselves so that the ambush is not compromised. They must exercise patience and self-discipline by remaining still and quiet while waiting for the target to appear. They have to forego smoking, endure insect bites and thirst in silence; resist the desire to sleep, ease cramped muscles, and perform normal body functions. When the enemy appears, they must resist the temptation to open fire before the signal is given. The patrol leader must effectively control all units of the ambush force. Control is most critical at the time the enemy approaches the killing zone. Control measures must provide for—

- Early warning of enemy approach.
- Fire control.
- Initiation of appropriate action if the ambush is prematurely detected.
- Timely and orderly withdrawal of the ambush force from the ambush site and movement to the objective rally point.

(8) Suitable Objective Rally Point. The objective rally point must be easily accessible to all personnel in the ambush force. It must be located far enough from the ambush site so that it will not be overrun if the enemy successfully assaults the ambush. Situation permitting, each man walks the route from his ambush position to the objective rally point. If a night ambush, he must be able to follow his route in the dark. After the ambush has been executed, and the search of the killing zone completed, the patrol withdraws quickly but quietly on the patrol leader's signal. If the ambush was not successful and the patrol is pursued, withdrawal may be by bounds. The last group may arm mines, previously placed along the withdrawal route, to further delay pursuit.

e. Execution of an Ambush. The manner in which the patrol executes an ambush depends on whether the purpose is to kill or harass the enemy.

(1) When the primary purpose of the patrol is to kill the enemy, the area is sealed off with the security teams. Maximum damage is inflicted with demolitions, command-detonated claymore mines, automatic rifle and machine gun fire, and antiarmor weapons. When these fires are ceased or shifted, the patrol launches a violent assault into the killing zone. The assault unit then provides security while designated teams kill or capture personnel, search bodies for items of intelligence value, and destroy vehicles and equipment. On the patrol leader's signal, all units withdraw to the objective rally point, reform the patrol, and move out quickly.

(2) When the primary purpose is harassment, the patrol seals off the area to prevent reinforcement and enemy escape. Maximum damage is inflicted with demolitions and automatic weapons fire. The patrol delivers a high volume of fire for a short time and then withdraws quickly and quietly. The patrol does not assault, except by fire, and avoids physical contact. The patrol avoids being seen by the enemy.

(3) When the patrol's primary purpose is to obtain supplies or capture equipment, security units seal off the area. Weapons and demolitions are used to disable vehicles, but not destroy them. The assault unit must use care to ensure its fire does not damage the desired supplies or equipment. Designated teams secure the desired supplies or equipment. Other teams then destroy vehicles and equipment not needed by the patrol.

f. Miscellaneous Ambush Techniques

(1) Normally, the ambush patrol will be deployed along a trail or route which is either known to be used by the enemy, or likely to be used. The enemy is permitted to pass by the center of the ambush force so the attack can be made from the rear. One or two men should be posted well forward and to the rear along the route, to prevent the enemy from escaping. All fires should be delivered simultaneously on a prearranged signal.

(2) It is important to remember that an ambush should have four distinct signals—one to open fire (with an alternate signal utilized at the same time as the primary), a signal to cease or shift fires, a signal to assault or search the killing zone, and a signal to withdraw.

(3) The signal to open fire should meet two criteria. First, it should be the firing of a weapon that will kill the enemy. Second, it should be a weapon that will shock the enemy and throw him into a state of confusion. An excellent primary signal is a command-detonated claymore mine fired by the patrol leader. The alternate signal should be a machine gun or an automatic weapon.

8507. Security Patrols

In a static situation, security patrols prevent the enemy from infiltrating the area, detect and kill or capture infiltrators, and prevent surprise attack. They protect a moving unit by screening the flanks, the areas through which the unit will pass, and the route over which the unit will travel.

a. Task Organization and Equipment. Patrol members are heavily armed and equipped to handle themselves in a meeting engagement with the enemy. They carry an ample supply of hand grenades and flares. The patrol must be equipped with reliable radio communications in order to report information over whatever distance from the parent unit it is employed. Pop-up flare signals are used as secondary means of communication.

b. Planning. A security patrol must not be viewed as a simple matter of following a designated route, calling in checkpoints upon arrival, and returning to friendly lines/area via a different route. The patrol must be well planned and rehearsed. Units, teams, and personnel must be designated and assigned specific missions. Its route must keep it within proximity of friendly positions, so that it can be supported with both organic and nonorganic weapons and reinforced, if necessary. Communications must always be maintained between the patrol and its parent unit. In relatively static positions, patrol routes and times must vary; routine patterns must be avoided.

c. Actions at the Objective. A series of objectives is selected covering the area over which the patrol is to move. These objectives may double as checkpoints. Actions at each objective are planned much in the same manner as are actions at danger areas.

Section VI. Information and Reports

8601. General

The patrol leader and every member of all patrols must be trained in observing and accurately reporting their observations. The patrol leader should have all members of his patrol signal or immediately report to him any information they obtain. These reports should not be restricted to information about the enemy, but should include any information about the terrain, such as newly discovered roads, trails, swamps, and streams. The patrol leader includes all information in his report to the officer dispatching the patrol.

8602. Sending Information

The patrol leader is informed if messages are to be sent back and what means of communication are to be used. Messages may be oral or written. They must be accurate, clear, and complete, and answer the questions, what, where, and when.

a. Oral Messages. An oral message must be simple, brief, and not contain numbers or names.

b. Written Messages. Written messages must contain facts, not the patrol leader's opinions. Information about the enemy should include strength, armament, activity, direction of movement, the patrol's location when the observation was made, and the time the enemy was observed. An overlay or sketch should accompany the message if deemed necessary.

c. Messengers. If a patrol leader must communicate a message of great importance and cannot do so by radio, he should dispatch two messengers, each taking a different route, to increase the possibility of having the information reach the person for whom it is intended. A messenger is given exact instructions as to where the information is to be delivered and what route is to be taken. If the message is oral, the patrol leader has the messenger repeat the message back to him before departing. Any information the messenger obtains along the way must also be reported at the time the message is delivered. If delayed or lost, he should show the message to an officer, if possible, and ask his advice. Messengers must be given all practicable assistance. If in danger of capture, the messenger destroys the message.

d. Use of Radio and Other Means. If the patrol is provided with a radio, a definite schedule for checking in must be established by the parent unit prior to the patrol's departure. The patrol leader must take every precaution to ensure that codes and copies of messages are not lost or captured. Radio transmissions from the patrol should be infrequent and short, and the patrol should leave the area immediately after transmitting to reduce the possibility of being detected by enemy direction finding equipment. Pyrotechnics (flares, colored smoke, etc.) and air panels may also be used by the patrol to communicate simple signals and information.

e. SALUTE Report. In order to keep messages brief, accurate, and complete, the SALUTE report should be used to report information, written or oral. Appendix F provides information on the SALUTE report.

8603. Captured Documents

Every patrol should search enemy dead and installations for papers, maps, messages, orders, diaries, and codes, after first ensuring that they are not boobytrapped. All captured documents are turned in to the patrol leader who turns them in when he makes his report. The documents should be marked as to time and place of capture.

8604. Patrol Report

Every patrol makes a report when the patrol returns. This report is made at a debriefing which is attended by the patrol leader and all patrol members. Unless otherwise directed, the report is made to the person ordering the patrol. If the situation permits, the report is written and supported by overlays and/or sketches. The patrol leader's report should be a complete account of everything of military importance observed or encountered by the patrol while on the mission. (See fiig. 8-3.)

8605. Patrol Critique

After the patrol has rested and been fed, the patrol leader should hold a critique. Constructive criticism is offered by the patrol leader and, in turn, by members of the patrol. It is an excellent time to prepare for future patrols by going over lessons learned as a result of the patrol. At the conclusion of the critique, the patrol leader should let his men know what their patrol accomplished.

PATROL REPORT

(OMIT HEADING(S) NOT APPLICABLE)

(DESIGNATION OF PATROL)   (DATE)

TO: ....

MAPS: ....

A. SIZE AND COMPOSITION OF PATROL

B. TASK

C. TIME OF DEPARTURE

D. TIME OF RETURN

E. ROUTES (OUT AND BACK)

F. TERRAIN ← DESCRIPTION OF THE TERRAIN – DRY, SWAMPY, JUNGLE, THICKLY WOODED, HIGH BRUSH, ROCKY, DEEPNESS OF RAVINES AND DRAWS; CONDITION OF BRIDGES AS TO TYPE, SIZE, AND STRENGTH; EFFECT ON ARMOR AND WHEELED VEHICLES.

G. ENEMY ← STRENGTH, DISPOSITION, CONDITION OF DEFENSE, EQUIPMENT, WEAPONS, ATTITUDE, MORALE, EXACT LOCATION, MOVEMENTS, AND ANY SHIFT IN DISPOSITION. TIME ACTIVITY WAS OBSERVED; COORDINATES WHERE ACTIVITY OCCURRED.

H. ANY MAP CORRECTIONS

J. MISCELLANEOUS INFORMATION ← INCLUDING ASPECTS OF NUCLEAR, BIOLOGICAL, AND CHEMICAL WARFARE.

K. RESULTS OF ENCOUNTERS WITH ENEMY ← ENEMY PRISONERS AND DISPOSITION, IDENTIFICATIONS, ENEMY CASUALTIES, CAPTURED DOCUMENTS AND EQUIPMENT.

L. CONDITION OF PATROL INCLUDING DISPOSITION OF ANY DEAD OR WOUNDED

M. CONCLUSIONS AND RECOMMENDATIONS ← INCLUDING TO WHAT EXTENT THE TASK WAS ACCOMPLISHED AND RECOMMENDATIONS AS TO PATROL EQUIPMENT AND TACTICS.

_____ _____ _____
SIGNATURE   GRADE/RANK   ORGANIZATION/UNIT OF PATROL LEADER

N. ADDITIONAL REMARKS BY INTERROGATOR

_____ _____ _____
SIGNATURE   GRADE/RANK   ORGANIZATION/UNIT OF INTERROGATOR   TIME

O. DISTRIBUTION

Figure 8-3. Patrol Report.

Chapter 9

Special Tactics and Techniques

Section I. Military Operations on Urbanized Terrain

9101. General

Fighting in urban areas is characterized by close combat, limited fields of fire and observation, difficulty in control of troops, and canalization of vehicular movement. The defender can use urban areas to canalize and stop an enemy attack. Buildings normally provide both the attacker and defender with good protection from both direct and indirect fire. They also provide covered and concealed routes that can be used to move troops and supplies. Fighting is at close range, and the outcome depends largely upon the initiative and aggressive leadership of small-unit leaders. In many cases, squads and fire teams operate independently in their assigned zones of action. The attacker must isolate and clear the parts of the urban area that he moves through. This causes him to lose the momentum of his attack. Urban areas must be attacked when they—

- Cannot be bypassed.
- Contain a key terrain feature (bridge, road junction).
- Must be returned to friendly control.
- Are needed for future operations.
- Are held by an enemy force that must be destroyed.

Military operations on urbanized terrain (MOUT) includes all military actions that are planned and conducted on a terrain complex where manmade construction impacts on the tactical options available to the commander.

Urbanization is characterized by changes in land usage and the spread of manmade features across natural terrain. There are four categories of urban terrain; strip areas, villages, towns and small cities, and large cities with associated urban sprawl.

The full range of manmade and natural terrain features exists on the urban battlefield. One end of the scale consists of small built-up areas dispersed along lines of communications. The force attacking through such an area may be able to bypass or easily sweep aside any resistance without losing the momentum of attack. The other end of this spectrum consists of a complex urban area which totally involves maneuver units in building-by-building, floor-by-floor, combat.

9102. Structural Classification

There are generally two types of structures; frameless and framed.

a. Frameless. (See fig. 9-1.) Frameless structures are those structures wherein the mass of the exterior walls performs the principal load-bearing functions of supporting the following:

- Dead-weight of roofs, floors, and ceilings.
- Weight of furnishings and occupants.

Building materials for frameless structures are mud, stone, brick, cement building blocks, and reinforced concrete. Wall thickness varies with material and building height. Frameless structures have much thicker walls than framed structures and are therefore more resistant to penetration by projectiles. Frameless buildings are usually restricted in the amount of area devoted to doors and windows.

Windows are usually in vertical alignment to maintain integrity of the load-bearing walls. Window dimensions may increase with higher placement in the building facade. The subtypes of frameless buildings vary with function, age, availability, and cost of building materials. Older institutional buildings, such as churches, are frequently made of stone. Reinforced concrete is the principal material for wall-and-slab structures (apartments and hotels), and in structures used for commercial and industrial purposes. Frameless structures are widely distributed. Brick structures, the most common type, dominate urban areas (except in the relatively few parts of the world where wood-framed houses are common). Close set brick structures,

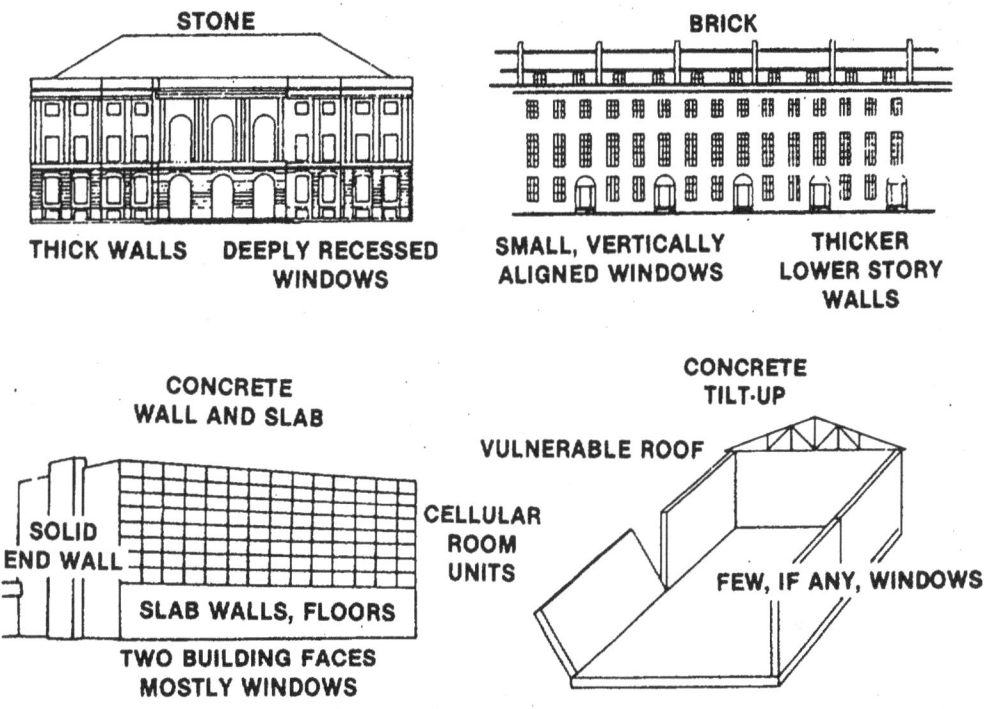

Figure 9-1. Frameless Buildings.

up to five stories in height, are located on relatively narrow streets and form a hard shock-absorbing protective zone for the inner city. The volume of rubble produced by their complete or partial demolition provides countless defensive positions, and narrow streets are easily blocked to impede traffic. In some cities, brick buidings have been replaced by concrete and steel structures in civic reconstruction projects. Approximately two-thirds of the world's total buildings are frameless. Brick structures account for nearly 62 percent of all buildings.

b. Framed. (See fig. 9-2.) This category is typified by a skeletal structure of columns and beams which support both vertical and horizontal loads. Exterior walls are nonload-bearing. Without the impediment of load-bearing walls, large open interior spaces, which offer little protection to defenders, are possible; the only available refuge is the central core of reinforced concrete present in many of these buildings (e.g., an elevator shaft). Multi-storied, steel and concrete framed structures have an importance far beyond their one-third contribution to the world's total buildings. They occupy the valuable core areas of cities, and as seats of economic and political power, may have a high degree of military significance.

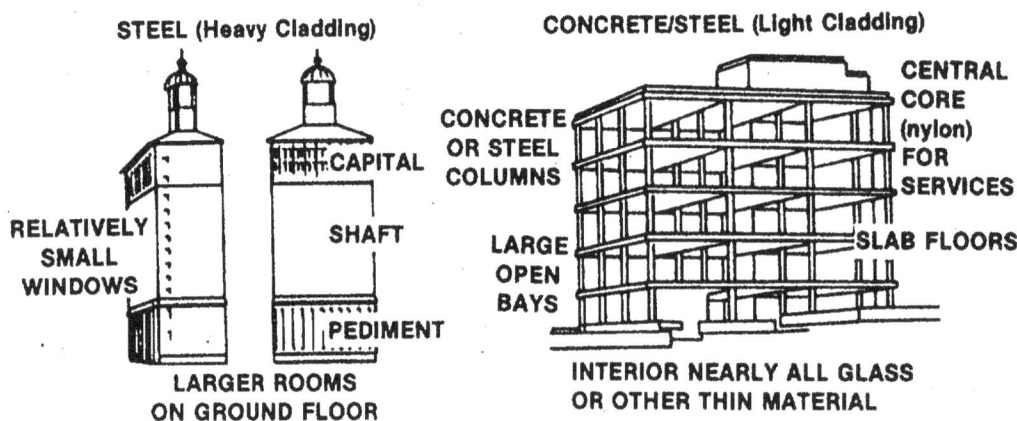

Figure 9-2. Framed Buildings.

9103. Tactical Considerations

A built-up area creates special tactical problems which must be considered by both the attacker and defender. Factors to be considered are:

  a. Cover and concealment are available to both the attacker and defender.

  b. Streets and alleys that provide the easiest routes of movement also provide ready-made fire lanes and killing zones.

  c. Observation and fields of fire are limited, except along streets and alleys.

  d. Restriction and canalization are imposed on vehicular movement. Both wheeled and tracked vehicles must be given close-in protection by infantry, due to their vulnerability to close range attack. Tanks lose much of the effect of their main armament when encountering enemy troops located in the upper stories of tall buildings or in basements. This is due to the main gun's limited range of elevation and depression at close quarters.

  e. The close proximity of assaulting and defending forces may restrict the use of mortars, artillery, and air support.

  f. High-angle fire weapons such as mortars become more important since they can place fire into the defiladed areas between buildings.

g. To a degree, walls and roofs of buildings will neutralize the fragmentation effects of artillery, mortars, and air-delivered bombs and rockets; however, the number of casualties caused by the blast effects of these weapons is increased due to collapsing roofs, walls, and flying debris.

h. The use of supporting arms by the attacker will create rubble which will afford the defender increased cover and concealment and serve to block armored avenues of approach.

i. With respect to observation and fields of fire, the possession and control of upper floors and rooftops offer an advantage to both attacker and defender.

j. Sound magnification and the echoes created by explosions and the firing of weapons will increase the difficulty of detecting enemy weapons. Dust and smoke from explosions and fires will reduce the ability to observe friendly fire and to detect and locate hostile fire.

k. Movement in the open is restricted. Speed, aggressiveness, and cunning are necessary to ensure the accomplishment of the mission.

l. Radio communications will be affected by the interference created by buildings.

m. Smoke can effectively provide concealment, limit observation, and facilitate deception and surprise. Smoke will remain effective for longer periods since it will not dissipate as quickly as in open terrain.

n. Tripod-mounted machine guns are employed to deliver accurate fire down streets and alleys even when the gunner's observation is obscured by smoke.

o. Incendiaries offer a quick means of dislodging a defender, but must not create an obstacle (burning buildings) to the advance.

p. Normally, the rifle company is the lowest tactical unit to maintain a reserve. Due to the decentralized nature of fighting in built-up areas, the rifle platoon is authorized to designate a reserve unit, normally a rifle squad.

9104. Phases of Attacking a Built-Up Area

a. In attacking a built-up area, the plan will normally be executed in three phases:

(1) Phase I. Phase I is designed to isolate the objective by controlling avenues of approach into and out of the built-up area. Key terrain outside the built-up area is secured. From here, supporting arms are employed to support the maneuver force's entrance into the built-up area and the step-by-step seizure of objectives; supporting fires also help to prevent enemy reinforcement of or escape from the area.

(2) Phase II. Phase II consists of an assault to rupture the defenses and secure a foothold on the perimeter of the built-up area. From here attacks to clear the area may be launched. An envelopment (assaulting defensive weaknesses on the flanks or rear of the built-up area) is preferred; however, a penetration may be required.

(3) Phase III. Phase III is predominantly a clearing action which consists of a systematic building-by-building, block-by-block advance through the built-up area. This phase is characterized by decentralized, small-unit actions, and requires detailed planning to offset the difficulties of control.

b. The attack does not stop or slow down between Phases II and III. There may be actually no distinction between the two. As each unit secures its foothold in the near edge of the built-up area, it immediately begins to displace its reserve and supporting weapons into the foothold area to support the third phase of the attack.

9105. Organization of the Rifle Squad

The rifle squad will fight as part of the rifle platoon in this type of operation. The rifle platoon in its assault of a built-up area will normally adopt the vee formation. The reserve squad will be positioned and displaced as required to protect the flanks and rear of the attacking squads. It should be prepared to assume the mission of either attacking squad on order or respond to a new mission if so ordered. Each of the two attacking squads will be formed into covering parties and search parties. The squad leader and any supporting weapons, such as SMAWs, will normally be positioned with the covering party.

a. Covering Party. The covering party, consisting of two fire teams, protects and assists the advance of the other fire team assigned as the search party. The covering party places fire on the building to be entered by the search party, on adjacent buildings, and establishes killing zones around the building being searched, in order to prevent enemy reinforcements from entering. The covering party displaces on order of the squad leader, under the protection of the fire of the search party, once the search party has secured the objective.

b. Search Party. The search party is further divided into a search team and a covering team. The search team (the rifleman and assistant automatic rifleman) makes the initial entry into the building under attack and clears the room. The covering team (fire team leader and automatic rifleman) enter the building after the search team has cleared the room. The covering team positions itself to cover the search team and to provide grenade launcher and automatic rifle fire as required. Areas of the building not yet cleared are kept under constant surveillance by the covering team. When clearing a building, the search team will search every space within the building to ensure it is clear.

9106. Search Party Procedures

a. If possible, buildings are cleared from the top down. There are three methods of entering and clearing buildings.

(1) Entry on Uppermost Level. Entrance through the upper part of a building is preferable, as it is easier to work down than up. An enemy cornered on the top floor of a building may fight desperately, but an enemy forced down to the ground level is likely to withdraw from the building, thus exposing himself to fire from the covering party and supporting machine guns.

(2) Entry on Middle Floor. If entry on the top floor is not possible, entry should be made at the highest possible point. The point of entry is cleared first. Clear the upper floors and roof, then work down.

(3) Entry on Ground Floor. When entering at the ground level, it is wise to use demolitions, SMAWs, light assault weapons (LAWs), tanks, or other weapons which produce similar effects, to blast an entryway instead of using windows or doors. Doors and windows located on the ground level are probably covered by enemy fire and may be boobytrapped.

b. A squad attacking in a built-up area is assigned to clear buildings within the assigned platoon frontage. The squad leader identifies the building to be searched and directs the covering party to deliver fire into the building. On the squad leader's order, the fires of the covering party are shifted from the point of entry. The search party then enters the building to be searched.

(1) If it is a single story building, a member of the search party approaches the entryway, takes appropriate cover by shielding himself against the wall of the building, and throws a fragmentation grenade through the entryway. He then hits the deck, diving away from the entryway, laying flat until the grenade explodes.

(2) From a distance the M-203, firing the HE or HE dual-purpose round can be used instead of a fragmentation grenade. The M-203 round(s) can be fired by either the covering party or the covering team, at the squad leader's discretion.

(3) Taking advantage of the shock effect of the grenade, one member of the search team enters the building and takes a position from which he can cover the entire room.

(4) The second member of the search team enters and searches the room.

(5) When the first room has been searched, the search team sounds off, **CLEAR,** and is joined by the covering team.

(a) Before entering the room, the fire team leader sounds off, **COMING IN**.

(b) After entering the room with the automatic rifleman, the fire team leader takes a position from which he can direct the search.

(6) Inside the building, voice communications are used.

(a) As each room is cleared, the search team sounds off, **CLEAR.**

(b) When entering or leaving a room which has been searched, sound off, **COMING IN** or, **COMING OUT.**

(c) When using stairs leading to areas which have been cleared, **COMING UP** or, **COMING DOWN** will be sounded.

(d) The use of voice communications prevents members of the search party from being surprised and assists the fire team leader in controlling the search.

(7) When the search team makes entries into subsequent rooms, the entry will normally be preceded by bursts of rifle fire. If the interior walls of the building are constructed of heavy (brick or concrete) material, fragmentation grenades may be used.

(8) The entry and searching process continues until the entire building is cleared.

(9) When the building is cleared, the fire team leader signals the squad leader and marks the building according to a prearranged code. The squad leader moves the covering party forward, protected by the search party.

9107. Techniques Employed

Combat in a built-up area requires the employment of techniques peculiar to this particular operation. Using these techniques enhances coordination and ensures accomplishment of assigned missions with minimum casualties.

a. The attack is conducted systematically, block-by-block and section-by-section. This system reduces the possibility of enemy resistance remaining active in the rear of advancing units.

b. Coordination of movement between individuals and units is of utmost importance so that proper covering fire and observation can be provided. Additionally, the establishment of killing zones down streets and across open areas requires close coordination between units. Whenever streets or open areas are crossed, the squad leader must coordinate the lifting or shifting of friendly supporting fires from that street or open area.

c. Built-up areas are ideal for the use of boobytraps. Certain steps may be taken to avoid them or lessen their effectiveness.

(1) Mark entrances into buildings and rooms used by the search team. The covering team should use these entrances, since they are known to be clear.

(2) When moving on stairs, take two or three steps at a time. This reduces the chances of stepping on a boobytrapped step.

d. Proper individual movement is important, and the method of movement used by an individual or unit must be thought out in detail. One situation may call for speed and aggressiveness, while another may require stealth. The following movement techniques have been found effective in reducing friendly casualties.

(1) When moving along streets and alleys, select cover in advance. Hug walls and move from cover to cover. Keep away from the middle of streets. Above all, do not bunch up.

(2) Don't cross streets unless absolutely necessary. Coordinate street crossings with the covering team, covering party, and supporting machine guns. When crossing a street, move directly rather than diagonally. To limit exposure, teams cross together when possible but disperse again as soon as the street is crossed.

(3) Do not silhouette yourself when climbing over walls or piles of rubble or moving into entrances. Roll over walls and rubble and move through entrances quickly.

(4) The roof of a building is a good place to enter a building, but a Marine on a roof is vulnerable to enemy fire from all directions. When forcing an entrance from the roof, stay low, crawling if necessary, to avoid silhouetting yourself to the enemy.

e. Where it is an advantage, use smoke to cover the movement of units. Smoke normally lingers in a built-up area. When using smoke, remember that it can also obscure the observation of the friendly covering team and covering party. The employment of smoke must be a calculated decision.

f. Ladders, notched logs, drainpipes, and grappling hooks can be used to gain entry at the upper level of a building. Another effective technique involves the use of a pole. One member of the search team holds the bottom of the pole steady. The other Marine climbs the pole hand over hand, using his feet to walk up the side of the building. Regardless of the technique used to gain entry to an upper level, the entry must be done from a covered location. If there is no cover from enemy fire, do not try to climb the side of the building. If available, sewers, drainage ditches, and subways can be used as covered approaches.

g. CS grenades can be employed to rout enemy defenders. Use of CS must be well thought out as it affects both friendly and enemy personnel. The use of CS should be addressed in the platoon or squad leader's order prior to any attack.

h. The search party moves to the building on a covered and concealed route using smoke grenades or smoke pots for additional concealment. The search party enters the building at the highest level possible, because—

- The ground floor and basement are usually more strongly defended.
- Windows and doors on the ground floor and in the basement are usually boobytrapped.
- The roof of a building is normally weaker than the walls.
- It is easier to fight down stairs than up stairs.

i. Care must be taken in using hand grenades.

(1) Evaluate the construction of the exterior and interior walls of a building before using hand grenades. If the walls, especially interior walls, are of light construction, you can become a casualty of your own grenade. When using hand grenades inside a building, take cover behind sturdy furniture.

(2) Don't throw hand grenades upstairs.

(3) After releasing the safety lever (spoon), throw it into rooms vigorously. It will ricochet around the room and deny the enemy the opportunity to throw it back.

j. Rifle rounds will penetrate most interior walls. Be sure you know who is on the other side of a wall you fire at.

k. Extreme care must be taken in the employment of the M-203 when fighting in a built-up area. The following characteristics of the M-203 must be considered.

(1) The M-203 using the HE or HE airburst round is ideal for use by the covering team to deliver explosive ordnance into the upper levels of a building before the search team enters.

(2) The HEDP round can be used to blast holes in exterior walls of most framed buildings. It is not a suitable weapon for firing from one room to another as the majority of the round's effect will be directed at the wall it hits and beyond, rather than in the room itself.

(3) The fuzes used on all HE, HE airburst, and HEDP rounds require that the round travel a certain distance from the weapon's muzzle before arming. The arming distance on the M-551 fuze is 14 to 28 meters while the arming distance for the M-552 fuze is 3 meters. In most cases, normal room dimensions will preclude the use of HE, HE airburst, and HEDP rounds when clearing the interior of a building.

(4) Another disadvantage of firing HE, HE airburst, and HEDP rounds while inside of a building is that, even if sufficient distance exists to employ these rounds, it will be difficult to fire the weapon into a room and take cover before the round explodes. The effect of the round could injure or kill the Marine firing the M-203.

(5) The M-203 multi-projectile round can be employed by the search team.

l. Once the building is clear, it should be marked using a prearranged signal; for example, a chalk mark over the door, or a sheet hanging out of a window.

9108. How to Prepare a Building for Defense

a. A rifle platoon normally defends one to three buildings. This depends on the size, strength, and layout of the buildings. The rifle squad normally defends one entire building, two small buildings, or part of a large structure. Some considerations for the defense are:

(1) Protection. Reinforced concrete or brick buildings protect best. A reinforced cellar is good. Wooden buildings should be avoided.

(2) Dispersion. It is better to have positions in two mutually supporting buildings than in one building which may be bypassed.

(3) Concealment. Obvious positions, especially at the edge of an urban area, should be avoided.

(4) Covered Routes. These are used for movement and resupply. The best routes are through or behind buildings.

(5) Fire Hazard. Buildings which will burn easily should be avoided.

(6) Fields of Fire. Fighting positions should offer good all-around fields of fire.

(7) Time. Buildings which need a lot of preparation are undesirable when time is short.

(8) Observation. The building should permit observation into the adjacent sectors.

b. Once the squad leader is assigned a building or buildings to defend, he positions the attached weapons teams and his own fire teams. Machine guns and automatic rifles should be on ground floors to provide grazing fire. Antiarmor weapons and M-203s should be positioned on upper floors to take advantage of the longer ranges offered and enable them to fire at the tops and flanks of armored personnel carriers and tanks. Each squad should have a primary and supplementary fighting position for continuous all-around defense.

c. The squad's food, water, and ammunition may be stockpiled at each defensive position. If there is a fire hazard, fire fighting equipment should be positioned throughout the building. The floors should be covered with a layer of dirt. Phone lines should be laid through the buildings. Radio antennas can be hidden by placing them next to walls.

d. Buildings should be improved:

- Ceilings are reinforced.
- Gas and electricity are shut off.
- Glass is removed from windows.
- Man holes are cut or *blown* between rooms and through floors for covered routes to alternate and supplementary fighting positions.

e. Mines and obstacles are used to cover dead space and to keep the enemy from using streets, alleys, or rooftops. These obstacles should be covered by fire.

f. Window positions should be in the shadows, and not right at the window. Curtains, or a piece of cloth hung across a window, will hide a defender in a darkened room.

g. Fighting positions should be improved with sandbags or rubble and have overhead cover.

h. Doors, hallways, stairs, and windows that will not be used by the defenders are blocked or screened.

i. All firing positions are camouflaged. Dusty areas can be covered with blankets or wet down with water to keep dust from rising when weapons are fired. Burlap is better camouflage garnish than foilage.

j. If LAW, SMAW, and Dragon positions are located within the squad position, the following characteristics must be considered.

(1) Because of their backblast, Dragons and SMAWs need a floor area of at least 3½ by 4½ meters (12 by 15 feet) to the rear of the weapon. LAWs must have a clearance of at least 1.2 meters (4 feet) to the rear of the weapon.

(2) The room must have an open area (ventilation) of at least 2 square meters (21 square feet). An open doorway 1 by 2 meters (3 by 7 feet) will meet this requirement.

(3) All Marines in the room must wear earplugs.

(4) Positions of other Marines cannot be to the rear of the weapons.

Section II. Attack of Fortified Areas

9201. General

A fortified area contains permanent defensive works. These works consist of emplacements, field fortifications, obstacles, and personnel shelters. They are disposed laterally, in depth, and are mutually supporting. A trench or tunnel system may be included to afford covered movement. A fortified area is deliberately planned to deny access to an attacker.

a. Characteristics. Fortified areas differ in construction and physical layout; however, they all possess similar characteristics.

(1) **Strength**

(a) Emplacements and personnel shelters are constructed of reinforced concrete, steel, or heavy timbers and earth. The bulkheads and overhead may be up to 10 feet thick. This construction provides the defender with cover from indirect fire weapons, small arms, and limited protection from direct fire weapons.

(b) The area is usually prepared in advance of hostilities which permits the use of natural camouflage; however, artificial camouflage may be used.

(c) Each emplacement usually contains one or more automatic weapons.

(d) Emplacements are mutually supporting; one protects the other. To attack one emplacement, the attacker must pass through the sector of fire of one or more other emplacements.

(e) Each emplacement is protected by infantry occupying field fortifications positioned around the emplacement. These field fortifications may have overhead cover.

(f) Tunnels and communication trenches are normally used to link emplacements within the fortified position.

(g) Barbed wire and other obstacles are used extensively in order to restrict the attacker's movement and to channel him into the sectors of fire of automatic weapons.

(h) Mines and boobytraps are normally employed in fortified positions.

(i) Communication wire is laid deep underground, thus, providing a relatively secure means of communication.

(2) Weaknesses

(a) Placing automatic weapons in fixed emplacements restricts the gunner's observation and, generally, prevents the weapon from being moved to an alternate or supplementary position.

(b) Emplacements depend upon mutually supporting positions for all-around observation and fields of fire. When one emplacement is destroyed, observation and mutually supporting fires are reduced proportionately.

(c) Generally, the emplacements can withstand the effects of artillery and mortar fire. Artillery and mortar fire are effective against the field fortifications around the emplacements. If the field fortifications have overhead cover, point detonating and delay fuzed rounds can destroy or collapse the fortifications. If the field fortifications do not have overhead cover, proximity fuzed rounds can be employed. In either case, the enemy infantry in the field fortifications will be forced to seek more protected locations, thus, weakening the enemy defensive position.

(d) The weakest points of emplacements are embrasures, air vents, and doorways. They provide the attacker with an opening to employ grenades, rocket launchers, demolition charges, and small arms fire.

b. Tasks. Infantry units attacking a fortified position are organized into a base of fire and an assault unit. The base of fire provides covering fire for the assault unit which goes forward, attacks, and seizes an assigned objective within the enemy fortified position. Both the base of fire and assault unit are reinforced or supported by appropriate combat power. The cumulative effect of several small units simultaneously conducting coordinated attacks against a fortified position will result in the penetration and destruction of that position. The following tasks must be accomplished in order to successfully attack and destroy assigned objectives within a fortified position.

(1) Neutralize the enemy infantry occupying the field fortifications and protecting the emplacement to be attacked. Neutralization means not only suppressing the enemy fire but also the ability to observe avenues of approach. Artillery, mortars, and the base of fire provide neutralization fires.

(2) Neutralize the enemy automatic weapons fire coming from the emplacement being attacked and from any other emplacement whose sector of fire the assaulting unit must pass through. To do this, mutually supporting emplacements must be attacked simultaneously. The actual neutralization of automatic weapons in the emplacements is best accomplished by a tripod-mounted machine gun using a traversing and elevating mechanism to deliver a high volume of accurate fire directly into the embrasure of the emplacement. One machine gun team is employed for each emplacement. The team is best employed from a covered and concealed standoff location where it cannot be suppressed by enemy fire. The machine gun team may not necessarily be attached to the attacking unit, but may be assigned the mission to provide direct support for the attacking unit.

(3) Destruction of the enemy emplacement on the objective is a critical task. The emplacement's field of fire may cover the approach of the assault unit or that of an adjacent attacking unit. In any event, it must be destroyed. Emplacements can be destroyed from ranges up to 250 meters using the SMAW. If a SMAW is not available, a Dragon antitank guided missile may be used to destroy an emplacement. If no assault weapons are available, the emplacement must be assaulted and destroyed with demolitions or grenades.

(4) The assault unit advances under covering fire from the base of fire and supporting arms fire. The assault unit must possess sufficient combat power to accomplish the following tasks:

- Destroy, neutralize, or overcome barbed wire and other obstacles encountered while moving forward.

- Kill or capture enemy sentinels or security posts covering avenues of approach.

- Assault the enemy position, and kill or capture the enemy personnel in or around the destroyed emplacement and in the surrounding field fortifications.

- Establish a hasty defense to repel any enemy counterattack.

(5) Once the assault unit has seized the objective, the remainder of the attacking unit moves onto the objective and consolidates the position.

c. Role of the Rifle Squad. Normally, a rifle squad will participate in the attack on one emplacement as part of the platoon. The squad can be assigned as the base of fire or the assault unit.

d. Employment of the Base of Fire

(1) Organic Weapons

(a) Rifle. Rifle fires are directed against enemy troops occupying field fortifications protecting the emplacement.

(b) Automatic Rifle. Automatic rifle fire is directed at troops in field fortifications at the sustained rate. If a machine gun is not provided to neutralize the enemy automatic fire coming from the emplacement, or if the machine gun becomes inoperable, the automatic rifle should be assigned to neutralize the fire from the emplacement. In this case, fire is directed at the embrasure at the rapid rate. When fires from the emplacement are neutralized, automatic rifle fires are shifted to troops in field fortifications, and are fired at the sustained rate.

(c) M-203 Grenade Launcher. M-203 fire is directed against enemy troops in field fortifications.

(2) Supporting Weapons

(a) If available, the MK-19 and .50 caliber machine guns can be employed in conjunction with the base of fire. If employed, these weapons will be used to neutralize the enemy infantry in field fortifications. Normally, these weapons will be employed from a standoff distance and will directly support the attack, rather than being attached to the attacking unit.

(b) White phosphorus (WP) rounds, delivered from the company's 60mm mortars, can be used to screen the assault unit from enemy observation.

(c) As was previously discussed, artillery and mortar support is provided to neutralize the enemy occupying the field fortifications around the emplacement.

e. Employment of Weapons by the Assault Unit

(1) **Organic Weapons.** Squad organic weapons are employed by the assault unit in the same manner as they are employed in a daylight attack. (See par. 4303c.)

(2) **Munitions Available.** The following munitions are employed by the assault unit in reducing fortified positions.

   (a) **LAWs.** LAWs are used by the assault unit to destroy enemy field fortifications on the objective. LAWs can also be used to complete the destruction of the emplacement by firing into the embrasure from close range.

   (b) **Demolitions.** Demolitions are used extensively in fortified areas to breach obstacles and destroy emplacements. If combat engineers are not attached to the squad for the attack, squad members will be required to employ the demolitions.

   (c) **Fragmentation Grenades.** These are used against enemy infantry dug in around the emplacement. Additionally, fragmentation grenades can be thrown through the embrasure or down ventilation apertures of the emplacement.

   (d) **Smoke Grenades and Pyrotechnics.** These are used primarily for signaling between the units of the squad and from the squad to the platoon commander. They are used to signal that the assault is about to begin and that the base of fire should shift or cease. Smoke grenades may also be used to screen movements of the assault unit from enemy observation.

   (e) **White Phosphorus Grenade.** A WP grenade is normally used to screen the movement of the assault unit and to neutralize enemy personnel in field fortifications.

(3) **SMAW.** If the SMAW is employed with the assault unit, the assault unit leader selects the SMAW firing position. This position must not be closer than 10 meters from the emplacement being attacked as the round has an arming distance of 6 to 10 meters. He instructs the SMAW gunner to engage and destroy the emplacement as soon as possible or, if surprise

is desired, to open fire upon signal. Surprise in this case refers not only to the enemy in the emplacement being attacked, but also to other enemy positions around or which support the emplacement.

9202. Squad Assigned Mission of Seizing an Objective

a. General. The squad may be assigned the mission of seizing an objective within a fortified area by itself. If assigned this mission, the following considerations should be addressed.

(1) A machine gun team should be assigned to neutralize the automatic weapon in the emplacement.

(2) A SMAW or Dragon team should be attached to destroy the emplacement.

(3) The squad must be provided bangalore torpedoes and other demolitions to clear mines and obstacles, and satchel charges to destroy the emplacement or other field fortifications. Combat engineers may be attached to perform these tasks; however, squad members must be proficient in the employment of bangalore torpedoes, satchel charges, and other demolitions, in the event combat engineers are not available. The squad should also be provided the capability to employ artillery and mortars.

b. Planning and Coordination. The squad leader receives the attack order from his platoon commander. His squad, with the additional combat power provided, is assigned the mission of seizing or destroying a certain emplacement. The other squads are assigned to attack other emplacements simultaneously. Attacks are usually conducted on a narrow frontage in an attempt to effect a penetration.

c. Planning the Attack

(1) During his reconnaissance, the squad leader observes and plans the following:

- Location of the emplacement on the squad's objective.
- Position of the embrasures and enemy fields of fire.

- Surrounding field fortifications and obstacles.
- Location of other enemy emplacements supporting his assigned objective.
- A tentative assault position and final coordination line.
- A route to the assault position.
- A position for the base of fire.
- A firing position for LAWs and SMAWs.

(2) Based upon his reconnaissance, his mission, and the mission of the platoon, the squad leader plans his attack.

(3) The squad leader determines whether he will remain with the base of fire or move forward with the assault unit. If he is going to lead the assault unit, which is the preferred method, he designates the base of fire leader, who is normally the fire team leader of the fire team assigned to the base of fire. If for some reason he must remain with the base of fire, he will designate the senior fire team leader as the assault unit leader.

(4) Upon completing his plan, the squad leader conducts a rehearsal, if feasible. During the rehearsal, he checks his plan and ensures that all squad members and attachments know their duties.

d. Preparation Fires

(1) Bombardment of the area by air, mortars, artillery, and naval gunfire may precede the attack.

(2) The squad leader will position the base of fire and any weapons attached to the base of fire under cover of the preparatory fires. The base of fire position is as far forward as possible.

(3) The fire teams and attached weapons, comprising the assault unit, remain in the assembly area or are moved forward to the attack position.

(4) The squad leader attempts to destroy the emplacement using the SMAW from the base of fire position and other supporting fires. If this fails, he sends the assault unit forward with the SMAW team attached to attack and destroy the emplacement.

e. Movement to the Assault Position. The base of fire neutralizes the enemy in the emplacement by firing into the embrasure. Additionally, the enemy in the surrounding field fortifications are taken under fire. When fires from the emplacement and the surrounding field fortifications are neutralized, the assault unit moves forward to the assault position. Smoke should be used to screen movement to the assault position and during obstacle breeching.

(1) The assault unit breaches any obstacle it encounters and destroys enemy personnel who cannot be bypassed.

(2) The assault position should be to the flank of the emplacement to take advantage of the defender's restricted observation and limited field of fire.

f. Assault of the Emplacement. The assault is conducted in the following sequence.

(1) The assault unit leader signals the assault. The base of fire is shifted or lifted on this signal. Smoke should be used to cover the assault.

(2) If the emplacement has not already been destroyed, the SMAW, with the assault unit, is employed to destroy it.

(3) The assault unit assaults the objective as soon as supporting fires are shifted or ceased.

(a) Remaining enemy are destroyed with small arms fire, M-203 fire, and hand grenades.

(b) The emplacement is searched for intelligence material, tunnels, or other entrances.

(4) After the objective is seized, the squad leader moves his base of fire forward. The squad then consolidates the objective and prepares for any counterattack. The squad may be directed to support by fire another squad's attack, or it may be directed to continue its attack on another objective.

Section III. Tank-Infantry Coordination

9301. General

The tank is primarily an offensive weapon employed to carry the fight to the enemy. Maximum use is made of its firepower, mobility, speed, and shock action. Restricting the speed of tanks to the speed of infantry moving on foot is not desirable as it negates most of the tank's capabilities. The information in this section applies to those situations where the terrain and/or enemy antitank defenses restrict the movement of tanks.

   a. Infantry Responsibilities. A rifle platoon commander directs the initial type of protection to be afforded the tanks by his rifle squads. As the attack progresses, the rifle squad leader decides whether to protect the tank physically or by short- or long-range fire. His decision as to which type of protection to employ is based primarily on the terrain and situation.

   b. Tank Responsibilities. Tanks can provide accurate high velocity direct fire support to the infantry. As a secondary role, tanks can assist in the antitank defense. Tank unit commanders make recommendations as to method of tank employment based upon the situation, terrain, mission of supported unit, and tentative concept of operation.

9302. Tank Capabilities and Limitations

   a. Capabilities. Tanks have four general capabilities. They are:

   (1) Firepower. The M-1 tanks are equipped with a 120 mm high velocity main gun, cupola-mounted .50 caliber machine gun, loader's 7.62 mm machine gun, and a coaxially-mounted 7.62 mm machine gun.

   (2) Armor Protection. Tanks have sufficient armor to withstand small arms fire and shell fragments. This protection makes them an effective weapon in the assault.

   (3) Mobility. Tanks possess cross-country mobility which permits rapid concentration and maneuver. Tanks can span an 8-foot ditch, ford water four feet deep without special fording equipment, and are capable of climbing 3-foot obstacles.

(4) **Shock Action.** Firepower, armor protection, and mobility of the tank, when properly employed, provide shock action.

b. Limitations

(1) **Inherent.** Size, weight, noise, and limited observation are inherent limitations.

(2) **Natural Obstacles.** Terrain limits the employment of tanks. Deep mud, rocky or stumpy ground, dense woods, swamps, and extremely rough ground limit their use. Rolling terrain, where cross-country mobility can be exploited, is the best type terrain for tank operations.

(3) **Manmade Obstacles.** Manmade obstacles canalize tank movement and restrict their employment. Obstacles frequently encountered are tank ditches, tank traps, and roadblocks reinforced with antitank mines and overwatched by antitank weapons. Urban areas are also considered obstacles to the employment of tanks.

9303. Tank and Infantry Teamwork

There are two important areas in which tanks interact with an infantry squad: movement to contact and actions on contact (including attacks). In both instances, squad leaders working with tanks should avoid a mind-set involving rigid formations or notions of the tank's invulnerability.

a. Movement to Contact.
During movement to contact, the tank's long-range firepower provides protection against enemy armor and enemy infantry. The infantry provides protection for the tanks against short-range ambush by enemy infantry. Clearly, tanks and infantry complement one another.

(1) **Open Terrain.** In open terrain, tanks provide overwatch for the infantry as they move forward. The infantry, in turn, may provide overwatch for the tanks. See FMFM 6-11, *Tactical Fundamentals for Mechanized Operations* (under development), for a thorough discussion of the overwatch movement technique. (See fig. 9-3.)

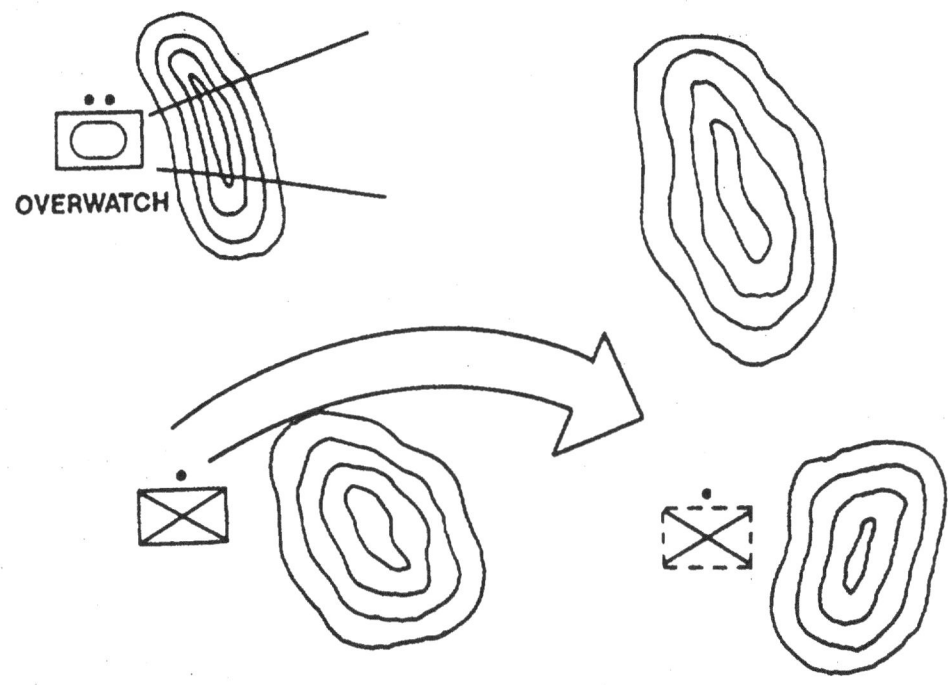

Figure 9-3. Open Terrain.

(2) Close Terrain. In heavily-wooded or jungle terrain, the infantry will normally lead the tanks. The tanks should provide overwatch, if possible. The infantry, in turn, is not only pursuing its mission but is providing a degree of close-in protection to the tanks. Close-in protection is **not** necessarily provided by standing next to the tank; close-in protection is provided by the infantry being able to engage the enemy before the enemy can fire on the tanks. The choice of formation, movement technique, and distance between units are METT-T-dependent. The principle of the infantry engaging the enemy before the enemy can engage the tanks is illustrated in figure 9-4.

b. Actions on Contact/Attack. The basic plan for a surprise encounter with the enemy and hasty attack or a well-planned deliberate attack is the same—fire and maneuver. Rigid formations and set-piece battle drills are to be avoided. Tanks may initially be used as a base-of-fire and may be directed to fire and move in concert with or separate from the infantry. Tanks should assault an enemy or move on to an objective **only** when enemy antiarmor weapons are suppressed.

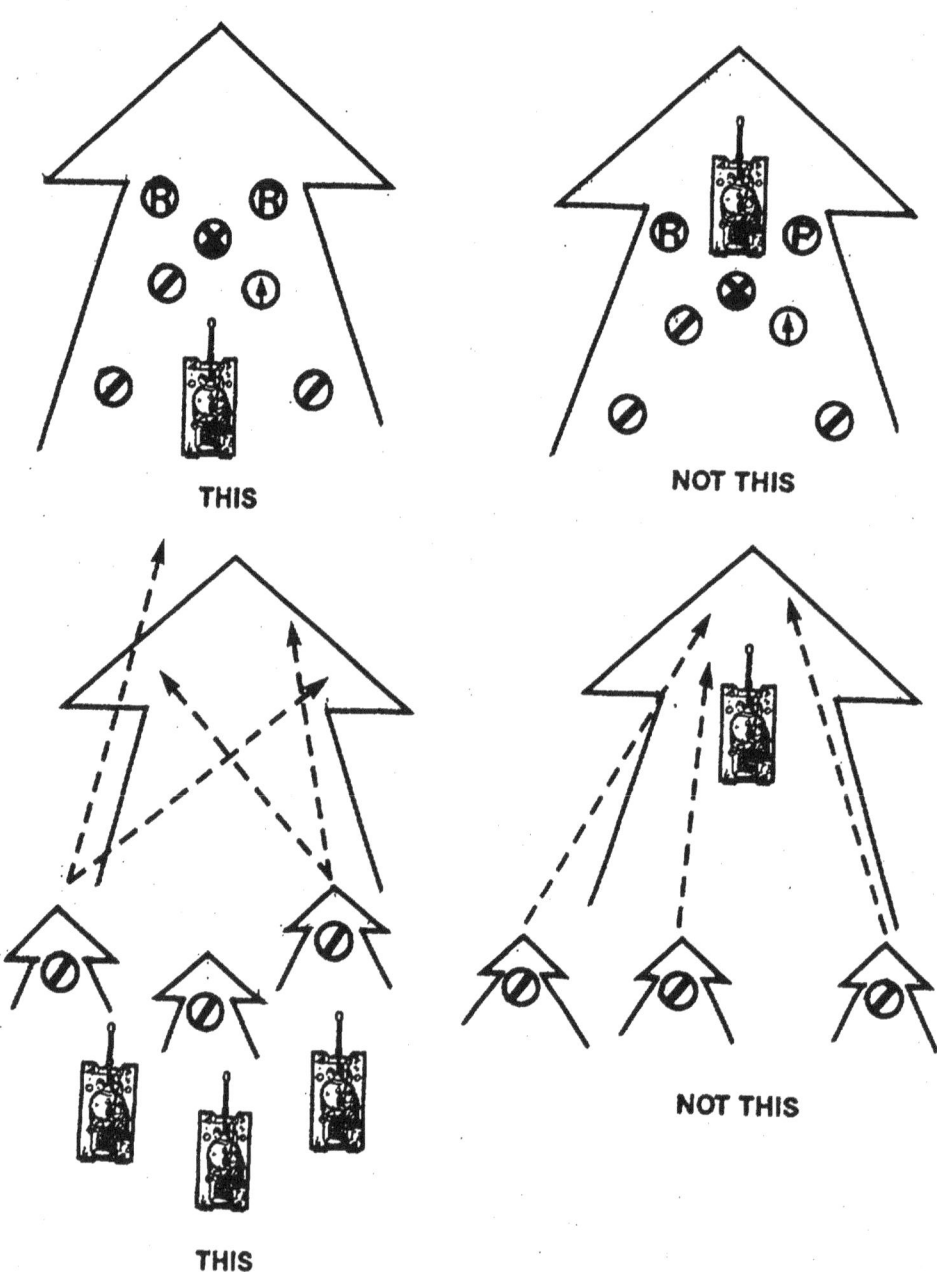

Figure 9-4. Close Terrain.

c. Communications. Communications between infantry and tanks may be accomplished by using agreed upon arm-and-hand signals, pyrotechnics, and radio. Backup communications and signals should always be planned.

d. Safety Considerations

(1) Do not move in front nor abreast of a tank when it is firing (muzzle blast area extends 20 meters on each flank).

(2) Tanks have a short turning radius. Stand clear of a moving or turning vehicle at all times.

(3) Stay clear of the immediate rear of the M-1 tank. The intense heat coming from its engine is dangerous.

(4) Infantry should keep clear of tanks being refueled or resupplied with ammunition.

Section IV. Mine Warfare and Demolitions

9401. Mine Warfare

Mine warfare is the employment of mines against an enemy and the countermeasures employed against the hostile use of mines. In both offensive and defensive combat, employment of minefields, covered by fire, helps a commander accomplish the mission.

The United States military *does not* employ non-self-destructing/deactivating antipersonnel/antitank landmines (NSDAPLs/NSDATLs) in accordance with the President's policy on landmines as outlined in Protocol II to the Convention on Prohibitions or Restrictions on the use of Certain Conventional Weapons Which May be Deemed to be Excessively Injurious or to Have Indiscriminate Effects (otherwise known as the CCW Convention). Self-destructing/deactivating antipersonnel/antitank landmines (SDAPLs/SDATLs) are found only in the family of scatterable mines (FASCAM). It is imperative that each Marine rifle squad be adequately trained to protect itself from enemy mines. Because of the relatively cheap price and simplicity of use, the landmine has become one of the most common threats to all military operations. Due to the broadened scale of enemy employment, mine warfare is no longer an exclusive engineer responsibility. Each Marine must be trained to gain confidence in dealing with both friendly and enemy mines.

a. Purpose of Minefields. A minefield is designed to counter armor, personnel, or a combination of the two. To prevent the enemy from bypassing the minefield, it is tied in with other obstacles, both natural and artificial, and/or supported by fire. Minefields are laid to—

- Delay the enemy.
- Canalize or guide the enemy to areas where fires may destroy him.
- Harass and/or demoralize the enemy.
- Supplement other weapons and obstacles.

b. Classification of Minefields. Minefields are classified according to their tactical purpose.

(1) Point Minefield. Point minefields are used to disorganize enemy forces and hinder their use of key areas. They include SDAPLs and SDATLs and may include antihandling devices. They are used to reinforce existing obstacles or rapidly block an enemy avenue of approach.

(2) Interdiction Minefield. Interdiction minefields are used in enemy-held areas to kill, disorganize, and disrupt lines of communications and command and control facilities. The squad will not normally have contact with interdiction minefields.

(3) Phony Minefield. A phony minefield is an area of ground used to simulate a live minefield and deceive the enemy. Phony minefields can supplement or extend live minefields and are used when time, effort, or material for live minefields is limited. Squad members may be called upon to assist in emplacing a phony minefield.

c. Landmines. Landmines consist of a high explosive charge contained in a metallic or nonmetallic casing fitted with either a fuze and/or a firing device for actuation by enemy vehicles or personnel.

(1) Antipersonnel mines consist of a small amount of high explosive charge in a container fitted with a detonating fuze arranged for actuation by *pressure* or *release of pressure,* by *pull on a tripwire*, or by *release of tension*. The two general types of antipersonnel mines are the bounding fragmentation type and the blast type.

(2) Antitank mines consist of a high explosive charge in a metallic or nonmetallic case. Antitank mines require a pressure of 290 to 500 pounds to detonate them. Actuation of antitank mines may be by pressure, tilt-rod assembly, or magnetic influence detonation.

(3) FASCAM are air- or artillery-delivered mines. They can be either antipersonnel or antitank mines. FASCAM mines are sown on top of the ground. In addition to the actuation methods mentioned above, FASCAM mines may have an antidisturbance actuator, which will explode the mine if it is touched or moved.

d. Minefield Installation. Minefields may be emplaced by hand, mechanical means, or may be air/artillery delivered. In a hand-emplaced minefield, the

Marine rifle squad may install all or a portion of the mines, depending on the size of the minefield and time available.

e. Reporting and Recording Minefields. Reports are made on every minefield placed or at the assumption of control of a friendly minefield. A *scatterable minefield report* must be made when emplacing a FASCAM minefield. This report must include at a minimum the approving authority, the type of emplacing system, the target/obstacle number, type of mines, life cycle of the mines, the aim and corner points of the field, and the safety zone from the aim point.

A *transfer of minefield report* must be made upon relinquishing or assuming control of a minefield. This report must include map sheets, grid coordinates of the minefield, obstacle identification number, and transfer from-to. It is the relinquishing unit's responsibility to provide all relevant information on a minefield in a clear and precise format.

f. Detection of Mines

(1) Visual. Visual inspection is used to locate mines. Experience with the mine habits of a particular enemy is often a great aid. A careless or hurried enemy may leave indications such as disturbed soil, piles of stones, or debris from mine packaging. Pot holes, road patches, soft spots in road and shoulders, bypass routes, and bivouac areas are places that Marines can expect to find mines. Abandoned vehicles, felled tree trunks, and souvenir materials should be avoided because of the boobytrap danger.

(2) Probing. Probing is the method of detecting mines by penetrating the ground with a sharp instrument such as a bayonet, stiff wire, or mine probe. Probing is slow, but it is the surest and safest way to locate mines, particularly the small antipersonnel type such as the M-14. When probing, move on hands and knees with sleeves rolled up to increase sensitivity to contact with tripwires. Look and feel for tripwires and pressure prongs, then probe at an angle of 45 degrees every 2 to 6 inches on a 1-meter *left-to-right* front. When an object is touched, stop probing, remove the earth carefully, and determine what has been hit. The Marine should then mark the mine's location and probe a new row about 2 to 6 inches forward of the last row.

(3) Electrical Detection. Electrical detection, using mine detectors, is a method used to locate most types of mines. The detector is effective only if operated by a trained and experienced operator.

g. Removal of Mines

(1) Hand Lifting. Hand lifting is dangerous, but is used when secrecy must be maintained or exploding the mine would cause undesired damage. After uncovering the mine and neutralizing any devices on top of and around the sides of the mine, probe under the mine and feel to locate and neutralize any secondary fuzes. Carefully lift the mine from the hole and move it aside for disposal.

(2) Rope Removal. Rope removal is accomplished by attaching a 50-meter rope or wire to the exposed mine and pulling the mine from the hole, utilizing a covered position. Wait at least 30 seconds before moving up to remove fuzes, pull wires, and activate devices.

(3) Removal by Explosives. Blow in place; this is the safest way to destroy mines. A 1-pound block of explosives set on or beside the mine is the most efficient method. The bangalore torpedo is suitable for breaching minefields to provide foot lanes.

9402. Demolitions

Explosives are not dangerous if handled according to safety instructions. Military explosives are not sensitive to heat, shock, and friction as a result of normal usage and transportation. Blasting caps, electric and nonelectric, are extremely sensitive and dangerous and must be carried in the specially designed cap box.

a. Training.
Training in explosives is simple and easily retained by the average Marine. This training includes preparing and using explosives to destroy bunkers, caves, bridges, and roadblocks; priming and firing of shaped charges, bangalore torpedoes, and claymore (M-18) mines; and blasting entrances into buildings.

b. Characteristics of Military Explosives

(1) Relative insensitivity to heat, shock, and friction. Not easily detonated by small arms fire.

(2) High detonating velocity.

(3) High power per unit weight.

(4) Stable under all climatic conditions.

(5) Detonated by easily prepared primers.

(6) Suitability for underwater use.

(7) Size and shape convenient for usage.

c. Explosives

(1) TNT, because of its insensitivity, stability, and power, is universally accepted as the basis for rating other explosives. Its shattering effect is suitable for breaching concrete, steel, or timbers. It can be used for all combat demolition missions.

(2) C-4 (C-3), a plastic explosive developed for steel cutting, is stronger than TNT and is suitable for all frontline demolitions.

d. Prepared Charges

(1) Shaped charges are used to blast boreholes in concrete and steel and can penetrate bunkers and heavy tank armor.

(2) Bangalore torpedoes are used to clear foot paths through barbed wire systems or antipersonnel minefields. They can also be used as an ambush weapon.

e. Demolition Accessories

(1) **Detonating Cord (Primacord).** An explosive cord used to detonate a charge or a number of charges simultaneously. It is designed as a primer, and does not possess the power to be used as a working explosive such as TNT or C-4.

(2) Blasting Caps

(a) Nonelectric. A small metal tube closed at one end containing PETN (pantaerythritol tetranitrate). It is highly sensitive and must be carried in a cap box.

(b) Electric. Same as above, except for two leg wires for electrical connection. Its use requires a source of current such as a 10-cap blasting machine (hell box) or battery and a sufficient length of wire to stretch well clear of the charge. By preparing this system, except for completing the circuit, explosions may be remotely controlled.

(3) Time Fuze (Safety Fuze).
A fuse used to detonate nonelectric blasting caps.

Section V. Nuclear and Chemical Defense

9501. Introduction to Nuclear Defense

The introduction of nuclear weapons in modern warfare has placed greater responsibility on the unit leader. While nuclear fires produce casualties through blast, heat, and radiation, their effect depends upon many variables. Such variables include the size or yield of the weapon, height of burst (subsurface, surface, or airburst), distance from ground zero, and the protection afforded troops by fighting holes or armor. Marines with a basic knowledge of nuclear weapons and their effects can survive and still function as an effective part of a combat unit. Tests have proven that troops with adequate protection can operate within a matter of minutes in an area where a nuclear explosion has occurred.

a. Conventional and Nuclear Explosions. There are several basic differences between a nuclear and a high explosive detonation. First, a nuclear explosion may be many thousands or millions of times more powerful than that of the largest conventional weapon. Second, a fairly large portion of the energy of a nuclear explosion is in the form of heat and light or thermal radiation, which is capable of producing injury or starting fires at considerable distances from the point of detonation. Third, and probably the greatest difference, is the highly penetrating and harmful rays, called *initial nuclear radiation*. Finally, the substances left after the explosion are radioactive, giving off harmful radiation over an extended period of time. This is known as the *residual nuclear radiation* or *residual radioactivity*. It is these differences between the conventional and nuclear explosions that require special considerations.

b. Effects on Individuals. Casualties from nuclear fire result from blast, thermal radiation, and nuclear radiation effects.

(1) Blast. Injuries are caused by both direct and indirect blast effects. Direct effect injuries, such as ruptured eardrums and internal injuries, are the result of the very high pressure waves generated by the blast. Indirect effect injuries are caused by falling buildings; flying objects; scattered glass; and fires started from short circuits, overturned stoves, and ignited fuels. The highest percentage of blast injuries will normally be a result of indirect effects.

(2) Thermal Radiation and Other Burns. It has been estimated that approximately 30 percent of the deaths that occurred as a result of nuclear weapons were due to burns of one kind or another. Persons in the open within 2 miles of a medium-sized nuclear weapon will receive painful flash burns on exposed skin. Fires resulting from blast cause other burn injuries. Personnel looking directly at the nuclear explosion may receive eye damage which is usually temporary in nature. Flash and flame burns resulting from nuclear explosions are treated like other type burns.

(3) Nuclear Radiation. The damage done by nuclear radiation depends on the dosage received and the time of exposure. Exposure to nuclear radiation does not necessarily cause radiation sickness. It takes a large amount of nuclear radiation, either initial or residual, to seriously harm an individual. Normally, the effects of nuclear radiation are not noticed during or immediately after exposure. If a man receives an excessive amount of nuclear radiation, such symptoms as nausea, vomiting, and a feeling of weakness occur within a few hours. Bear in mind, however, that a person can have nausea, vomiting, and weakness and still continue his duties. Generally, these immediate effects do not require that the individual be evacuated. Except in cases of extreme overexposure, these effects soon disappear and may not occur again, depending upon the dose received.

c. Types of Bursts. Nuclear bursts are generally described as air, surface, and subsurface, depending upon where the explosion occurs in relation to the surface of the earth. The height of burst is important since it influences the amount and kind of damage that occurs.

(1) Airburst. An airburst is a nuclear explosion in which the fireball does not touch the surface of the earth. The greatest danger from this type of burst is from blast and heat. The primary nuclear radiation hazard is from initial nuclear radiation, but a residual hazard may exist out to several hundred meters from ground zero (further with larger yield weapons), and should be avoided or passed through rapidly if attack through or reconnaissance of this area is necessary.

(2) Surface Burst. A surface burst is one in which the fireball touches, but is not beneath, the surface of the earth. Damage from blast is less widespread, and damage from heat is approximately the same as for an airburst of the same size. About the same initial nuclear radiation is present. Residual nuclear radiation is created in the target area and may occur as fallout in downwind areas.

(3) Subsurface Burst. A subsurface burst is one in which the center of the fireball is beneath the surface of the earth. Most of the blast effect appears as ground or water shock waves. The majority of the heat and initial nuclear radiation is absorbed by the surrounding earth or water. However, considerable residual radiation is produced in the target area and later as fallout.

d. Individual Protection. The effects of a nuclear explosion may be divided into two broad categories, immediate and delayed. The immediate effects are those which occur within a few minutes of detonation and include blast and shock, thermal radiation, and initial nuclear radiation. The delayed effects are normally associated with the radioactivity present in fallout and neutron-induced activity. The early fallout from a surface burst will begin to reach the ground within a few minutes after the explosion at close-in locations, and later at greater distances from ground zero.

(1) Protection from Blast. Since the effects of blast are immediate, consider the individual in each of two circumstances; when adequate warning is given prior to the attack, and when no warning is forthcoming.

(a) When adequate warning is given, the individual has time to prepare his defenses. The fighting hole is good protection. It should be deep and strong with the cover well secured. Bunkers, fortified positions, and shelters are excellent protection from all effects of blast.

(b) When there is no prior warning, individual reaction is mandatory. Remember, the flash can be seen a few seconds before the blast wave arrives. In any case, a person should stay down for at least 90 seconds. Ditches, culverts, and hills offer good protection and as the distance from the ground zero increases, good protection can be obtained from walls, slight depressions in the ground, or anything that breaks the pattern of the wave.

(2) Protection from Thermal Radiation. Another of the immediate effects that requires reaction on the part of the individual is the thermal radiation that is emitted from the fireball. Thermal radiation has a line of sight effect, so protection from it is the same as for blast. If prior warning of a nuclear attack is received, preparation both for blast and thermal radiation should be complete. If a weapon goes off with no warning, drop flat on the ground or to the bottom of a fighting hole, keep your eyes closed, and protect exposed skin from heat rays as much as possible (keep hands and arms near or under your body and helmet on). This immediate reaction will minimize serious burns.

(3) Protection from Nuclear Radiation. The different effects of nuclear radiation require both immediate and delayed action. Initial nuclear radiation is spent in the first minute or two after the burst, so protection from it is the same as for thermal radiation and blast. Any material will afford some protection from initial radiation, but denser items are better. For example, earth is a better shield than water, and steel is better than concrete.

(4) Protection From Residual Radiation. Under conditions where the explosion takes place either on or beneath the surface, the resulting residual radiation hazard is high. Particles of water spray, dust, and other debris, which become radioactive through contact with the nuclear reaction of the weapon, contaminate large areas. Individual protection consists of avoiding the fallout particles. This may be accomplished in the field by covering fighting holes with earth, a shelter half, or poncho. Open food and water supplies should be destroyed if contaminated. Personnel and equipment should be checked and decontaminated if necessary.

e. Decontamination. At the small-unit level, decontamination is usually confined to personnel, equipment, and food.

(1) Personnel. Personnel decontamination should be accomplished as soon as the tactical situation permits.

(a) Normal Procedure. In rear areas, and when permitted in the tactical area, personnel bathe, using plenty of soap and water. Particular attention should be given to skin creases, hairy parts of the body, and the fingernails.

(b) Field Expedient Procedures. When the tactical situation prohibits normal procedures, field expedient procedures are used. Clothing should be removed and shaken vigorously downwind. Shrubbery can be used to brush radioactive particles from the clothing. Personnel should put on a protective mask or cover their nose and mouth with a damp cloth to prevent inhalation of the radioactive dust. Care must be taken to avoid secondary contamination of food or water supplies during the shaking of clothing. Personnel then wipe all exposed skin with a damp cloth and remove as much dust as possible from the hair and from under the fingernails. Personnel should bathe and change clothing when the tactical situation permits.

(2) **Equipment.** The squad may be required to decontaminate individual items of equipment. The decontamination of equipment may be accomplished by removal (brushing and washing, sealing, and aging). In some cases, brushing will reduce dry contamination to a permissible level. In most cases, washing will be adequate even though brushing has not been effective. When speed in the decontamination of equipment is important, brushing is performed first, followed by washing as time and circumstances permit.

(3) **Food.** Food and water that have become contaminated should be destroyed by burying.

9502. Chemical Defense

Chemical agents are efficient, simple to produce, and capable of killing or incapacitating in a matter of seconds. Prior training and indoctrination is necessary for survival and combat efficiency.

a. **Types of Chemical Agents.** Chemical agents are divided into the following general classifications:

(1) **Toxic Chemical Agents.** Chemical agents designed to kill or incapacitate are known as toxic agents. They may enter the body through the lungs or by contact with the skin. The current groups of toxic agents are:

(a) **Nerve Agents.** These agents are quick-acting. Entering the body through breathing or by skin contact, they are very rapid in action and produce immediate casualties.

(b) **Blister Agents.** These are chemical agents which injure the eyes, lungs, and burns the skin. Symptoms may appear immediately or as long as 36 hours after exposure.

(c) Blood Agents. These are rapid acting agents that deprive the body of oxygen. They must be inhaled to be effective, so the field protective mask affords complete protection.

(d) Choking Agents. These are delayed action casualty producers that are very damaging to the lungs. Low concentrations may be difficult to detect. The protective mask provides adequate defense.

(e) Incapacitating Agents. These chemicals will cause temporary casualties that may be extremely difficult to control. Symptoms of mental confusion may take several hours to appear and may last for several days. Restraint and immediate medical assistance is necessary.

(2) Irritant Chemical Agents. The irritant agents produce temporary irritating or disabling effects if they are inhaled or contact the eyes and skin. The standard agents of this group range from CN (tear agents) through chlorine.

(3) Smokes. Smokes are chemical agents used to deny observation or transmit a signal.

(a) Screening Smokes. Screening smokes are used to screen friendly movement or deny enemy observation.

(b) Signaling Smokes. Signaling smokes are special fuels containing dye which produce colored smokes.

(4) Flame Fuels and Incendiaries. The flame fuels are normally employed against personnel and incendiaries are used to destroy material.

(a) Flame Fuels. These agents consist of special blends of petroleum products, usually in thickened form. They are easily ignited and may be projected toward a target by a flame weapon or flame tank.

(b) Incendiaries. These agents consist of a combination of flammable substances that burn with an intense heat and cause combustion and/or ignition in most materials. The incendiaries are thermite and thermate. White phosphorus may be used as an incendiary or as a screening smoke.

b. Detection of Chemical Agents. The success of chemical defense depends to a great degree on the thoroughness of the training program conducted by the small-unit leader. This training must consider the various characteristics of the agents as stated above and the three phases of defensive operations; detection, protection, and decontamination.

(1) Intelligence Sources. Intelligence sources may warn of expected attacks. These reports are usually based on enemy preparations and capabilities.

(2) Individual Marines. Marines may be able to identify a chemical attack by the use of their physical senses. Agents may have characteristic odors, create a visible cloud, appear as droplets on vegetation, or be detectable only by early recognition of symptoms. Because of this variety and the risk of rapid casualty effects, an automatic masking procedure will be put into effect once chemical operations are initiated. Any suspicious occurrence (low flying aircraft, smoke screen, unaccountable liquid, unusual physical or mental symptoms) will be considered a potential threat and all Marines will mask. The situation can then be checked in comparative safety and the decision made to unmask or continue a protected posture as required.

(3) Special Equipment. There are several items of special equipment designed for the identification and detection of chemical agents. The squad should be basically familiar with this equipment and its uses.

(a) Paper, Chemical Agent, Detector, ABC-M-8. ABC-M-8 chemical agent detector paper is a component of the chemical agent detector kits. The sheets consist of paper impregnated with chemical compounds that vary color when in contact with V- or G-type nerve agents or blister (mustard) agents, in liquid form. This paper does not detect vapor and must touch the liquid agent to ensure a positive test. Because some solvents cause a change in the color of the paper, it is unreliable for determining the completeness of decontamination by the use of solvents. A color chart is included in the booklet to aid in interpreting the tests.

(b) Chemical Agent Detector Kit, M-256. This item is designed for company and larger sized units and provides the means of detecting and identifying vapor concentrations of most chemical agents. These devices are designed for rapid identification of agents but cannot be used as warning devices as test reactions may take several minutes to complete.

(c) Individual Chemical Agent Detector (ICAD). ICAD is a small battery-powered chemical agent sensor that attaches to outer clothing. This device provides the means for a squad-size unit to detect low vapor concentrations of nerve, blood, blister, and choking agents.

(d) Chemical Agent Monitor (CAM). CAM is a handheld chemical agent monitor designed to identify the presence of agent vapor and residue for decontamination operations. CAM operates by two push buttons for on/off and mode change for selection of either nerve or blister agent detection.

(e) Water Testing Kit, M-272. The M-272 is designed to test for several types of chemical agents in salt, brackish, fresh, or treated water regardless of waterborne or possible interfering substances in any combination.

c. Protection

(1) Protective Mask. When properly fitted, the protective mask protects against inhalation and facial contamination by toxic agents. This is the primary means of protection in chemical defense.

(2) Protective Clothing. Protective clothing is available for those persons required to enter or remain in a contaminated area for a length of time.

(3) Antidotes. Antidotes for blister and nerve agents are found in the protective mask carrier. Atropine is used as treatment for any exposure to a nerve agent. It may be given by medical personnel or self-administered by the individual. Effects should be obvious in 15 to 20 minutes.

d. Decontamination. Decontamination consists of washing away, neutralizing, or destroying the chemical agent. The squad should be familiar with two groups of decontaminants, the natural and standard items, plus the procedure for personnel decontamination.

(1) Natural. Natural decontaminants are those provided by nature.

(a) Weather may be used when time is not a factor in the use of equipment or terrain.

(b) Water may be used to flush or neutralize certain agents from the surface of equipment. Hot water produces the best results.

(c) Earth is used to seal in contamination or act as an absorbent. If earthmoving equipment is available, an area that is contaminated may be covered with about 4 inches of earth and then crossed without danger to personnel.

(d) Fire may be used to destroy or vaporize liquid agents, especially in grassy or wooded areas.

(2) Standard. There are certain chemical compounds that may be used to reduce the effectiveness of contamination.

(a) Bleach (STB) is a chlorinated lime that will neutralize most liquid agents.

(b) Slurry is a mixture of bleach and water designed for easier surface coverage.

(c) DS-2 is a special solution designed for use against blister and nerve agents.

(d) The M-258 decontamination kit contains two separate packets, marked DECON 1 and DECON 2, to decontaminate skin and selected personal equipment.

(3) Personnel Decontamination

(a) Use the M-258 personal decontaminating kit immediately upon attack. A thorough washdown with hot, soapy water and a change of clothes should follow use of the kit, when the situation permits.

(b) The M-291 skin decon kit consists of a wallet-like flexible carrying pouch containing six identical, hermetically-sealed, foil packets. Each packet contains a folded, nonwoven fiber applicator pad filled with decontaminant powder and an attached strap handle on one side. Use this kit in the same manner as the M-258.

Section VI. Guerilla Operations

9601. General

Operations against guerrilla forces are characterized by aggressive and carefully planned offensive actions. The rifle squad leader must be aware of the role his squad plays when engaged in such operations.

9602. Characteristics

a. The squad leader must know how guerrillas operate to defeat them. Guerrillas require the following:

(1) Base. An area with dispersed and alternate facilities to provide security, discourage pursuit, provide routes to alternate bases which are close to the area of operations, and has adequate routes for entry and exit.

(2) Supply. A source of food, weapons, ammunition, and equipment.

(3) Intelligence. Information which enables him to plan operations or evacuate his bases when endangered.

(4) Communications. Means of gaining timely information and passing instructions promptly.

b. Guerrillas are ruthless, cunning, and able fighters even though they may have limited and outdated equipment. They attack weak or unprotected units at night, during periods of reduced visibility, and when security is likely to be lax. Attacks are characterized by speed, surprise, and rapid withdrawal.

9603. Indoctrination for the Squad

The squad leader must indoctrinate the squad prior to taking part in operations against guerrillas. Stress is placed on the following:

a. Self-Discipline. An important part of combating guerrillas is self-discipline. Men are proud of the spiritual values, culture, and customs of their country. If members of the squad ignore or neglect these items, hatred of Marines and sympathy for guerrillas may result.

b. Security. Marines must be security conscious at all times. Two routine methods of providing security are safeguarding information and guarding installations. Loose talk, as well as careless performance of duty, endangers friendly forces.

c. Need for Intelligence. Successful operations against guerrilla forces depend upon accurate and timely information. Guerrillas are elusive and must be located and eliminated. All means of communication are used to ensure rapid transmission of messages in keeping with security.

9604. Establishing a Patrol Base

To cover the entire area of guerrilla operations, it is usually necessary to establish temporary patrol bases some distance from the parent bases. Temporary patrol bases are established by company or smaller units to include the squad and should not be occupied more than 24 hours except in an extreme emergency. In all situations, a patrol base is occupied the minimum time necessary to accomplish the purpose for which it is established.

a. Deception. A patrol base is secretly occupied. Secrecy is maintained by practicing deception techniques that are carefully planned. Deception plans should include the following considerations:

- If possible, the march to the base is conducted at night.
- The route selected avoids centers of population.
- If necessary, local inhabitants met by the patrol in remote areas are detained.
- Inhabitants of areas that cannot be avoided are deceived by marching in a direction which indicates that the patrol is moving to some other area.
- Scouts operate forward of the main body of the patrol.
- Bases are usually located beyond areas that are patrolled daily.

- If fires are necessary, smokeless fuel is burned.
- Normally, not more than one trail should lead into the base and it should be camouflaged and guarded.
- The base is occupied as quickly and quietly as possible.
- The route to the base is selected by use of photos, maps, and ground and aerial reconnaissance.
- If practical, the patrol leader makes an aerial reconnaissance.
- Terrain features that are easily identified are selected as checkpoints and rest breaks.
- Daily aerial and ground reconnaissance is continued.

b. Locating the Base

(1) It must be secret and secure. A patrol operating from a base unknown to the enemy increases the possibility of guerrilla contact. A secure base permits the troops to rest.

(2) The base must have facilities of terrain suited for the erection of adequate radio antennas.

(3) If it is anticipated that an air drop or a helicopter resupply will be required, the base should have a convenient drop zone or landing point.

(4) The base must allow men to sleep in comfort. Wet areas and steep slopes are avoided. Flat and dry ground that drains quickly affords the best location.

c. Sequence of Establishment. A suggested sequence for establishing a base in jungle or heavy woods is as follows:

(1) Leaving the Road or Trail. The jungle and heavy woods provide the best security from surprise and the best conditions for defense. Generally the best methods to use in leaving the trail or road are:

- Select the point to leave the trail or road.
- Maintain security while the column moves off the trail.
- Have troops at the end of the column camouflage the area where the exit was made from the trail.
- Continue movement until a suitable patrol base is reached.

(2) Occupation of a Patrol Base

(a) This occupation is based on a platoon of three squads, but the force may be larger or smaller.

(b) The patrol is halted at the last suitable position approximately 200 meters from the tentative patrol base location.

(c) Close-in security for the patrol is established and the patrol leader has his subordinate leaders join him to conduct a reconnaissance.

(d) Patrol leader moves to the tentative patrol base location and designates point of entry into patrol base location as 6 o'clock, then moves to and designates center of base as patrol headquarters.

(e) Subordinate leaders reconnoiter areas assigned by clock system for suitability and return to patrol leader upon completion.

(f) Patrol leader sends some men to bring the patrol forward.

(g) Patrol leaves line of march at right angles and enters base single file moving to center of base. Designated men remove signs of patrol's movement.

(h) Each leader peels off his unit and leads it to the left flank of his assigned sector.

(i) Each unit occupies its portion of the perimeter by moving clockwise to the left flank of next sector.

(j) Each unit then reconnoiters forward of its sector, with designated individuals, by moving a specified distance out from the left flank of the sector; moving clockwise to the right limit of the sector; and reentering at the right flank of his sector. They report indications of enemy or civilians, suitable observation and listening post positions, rallying points, and withdrawal routes.

(k) Patrol leader then designates rallying points, positions for observation posts and listening posts, and withdrawal routes.

(l) Each squad then puts out an observation post (day) and a listening post (night) in front of each sector, establishes communications, and commences base routine.

d. Base Alert. The critical periods for defending the base are at dawn and dusk. During these periods, the entire patrol remains in an alert status. The base alert serves the following purposes:

(1) Enables each man to see the disposition of troops and the nature of the ground to his front and flanks.

(2) Provides a definite cut off period for the change of routine. Beginning with evening alert, all movement and noise cease and lights are extinguished. After the morning alert, the daily routine begins.

(3) Enables the squad leaders to check details while all men are positioned. This will include a check on maintenance of weapons, equipment, ammunition, etc.

e. Alarm. The patrol must have a suitable alarm signal for the approach of either friendly or enemy troops. This signal should not sound foreign to the environmental area and must be detected only by patrol members.

f. Leaving a Base. Before leaving the base, all signs of occupation are removed. Any shelters are destroyed. The area is left to appear as though it had not been occupied.

9605. Establishment of Control Over Civil Populace

Rifle squads will assist in carrying out steps to reduce sympathy and civilian support for guerrillas by making visits to villages, enforcing troop discipline, and working closely with civil authorities. Every effort is made to deny guerrillas their source of supply, reinforcement, and recruiting.

Surprise attacks against guerrillas will encourage the populace to resist their operations. When friendly forces react effectively to guerrilla attacks, insecurity is developed among the enemy, while the population gains confidence in friendly forces. Raids and ambushes are conducted to keep guerrillas in a state of alarm for their security, to lower morale, to prevent rest, and to hinder their operations.

Rifle squads may assist civil populace in executing the following control measures:

- Establishing restricted areas.
- Enforcing curfews.
- Relocating villages and settlements.
- Controlling weapons.
- Denying food.
- Searching individuals, vehicles, and houses.
- Training self-defense units.
- Riot control.
- Registering civilians.
- Establishing roadblocks and checkpoints.

9606. Patrol Operations Against Guerrillas

Short- and long-range patrols perform reconnaissance and combat missions in operations against guerrillas. Successful execution of these missions requires consideration of the significant differences between conventional and guerrilla forces.

Guerrillas are difficult to identify because they frequently dress in civilian clothing. Although a distinctive insignia is sometimes worn, the carrying of a weapon is often the only means of identification.

Guerrillas are elusive and when threatened scatter by fading into the civilian population, thus making decisive combat difficult. Certain persons may be of value to patrols engaged in operations; e.g., persons speaking the language of the area, or native to the areas, who are sympathetic to our cause. They may be able to identify guerrillas, help locate guerrilla bases, and provide other timely information.

Rapid communications are important. Immediate transmission of information permits prompt, decisive action against guerrillas.

9607. Offensive Action Against Guerrillas

Aggressive action is required to destroy guerrillas. Offensive action is conducted by applying the tactics and techniques of regular offensive combat.

a. Purpose. The primary purpose of offensive action is the destruction of guerrilla forces. Other objectives include capture or destruction of guerrilla concentrations, headquarters, communication centers, and sources of supply.

b. Types of Offensive Operations. The most effective forms of offensive action may be classified as encirclement and attack.

(1) Encirclement. Encirclement is the most effective method of destroying guerrillas. A rifle squad participates in the encirclement as part of a larger force. The operation consists of a movement from assembly areas to a line of encirclement, occupation of the line of encirclement, offensive drives, and destruction of guerrillas within the area.

(2) Attacks. Although the encirclement is the most effective offensive maneuver used in combating guerrilla forces, it is often difficult to execute because of inadequate forces, the type of terrain involved, or limitations imposed by the situation. Pressure on the enemy must be maintained by attacks, patrols, and raids.

(a) Squads attack using normal offensive tactics. Surprise gives the attacker a marked advantage; therefore, every effort is made to move all units into position quickly and undetected. Squads normally use the hours of darkness to move into positions for the attack and maneuver to blocking positions which seal off escape routes. All types of offensive tactics are employed. Success of the attack is often determined by aggressiveness and speed.

(b) If possible, a double envelopment is used by the main body. This maneuver limits guerrilla withdrawal, disorganizes them, and creates shock effect. Units of the attacking forces close rapidly with the enemy. Squad leaders must realize that the objective is total destruction of the guerrillas, not merely seizing and holding ground.

9608. Attacking Guerrilla Houses

In planning an attack:

- Secrecy is essential. Relatives, sympathizers, or intimidated natives can warn the enemy of the patrol's approach.

- Location of the house and the nature of terrain surrounding it are determined by ground or aerial reconnaissance, sketches, photos, or guides.

- The patrol approaches and occupies its position during darkness.

- The patrol should be no larger than required to carry out the mission.

- Approach is quiet and cautious, using all available cover.

- Avenues of escape are covered physically or by fire.

- If the mission is to capture the occupants, and armed resistance is not expected, surround the house and approach it from all sides.

- If the mission is to attack the house, and armed resistance is expected, the patrol is located so that every side of the building is covered by fire.

Appendix A

Military Symbols

1. General. Military symbols are used to identify and distinguish particular military units, activities, or installations. Listed herein are some of the military symbols commonly used by the squad leader.

2. Colors. Colors are used to assist anyone reading a map or overlay. Rules for their use are as follows:

a. Blue or black is used for friendly forces.

b. Red is used for enemy. If red is not available, blue or black in double lines can be used.

c. Green is used to show minefields, demolitions, roadblocks, and other engineered obstacle activities of both friendly and enemy forces. It will not be used for any other military activity. If not available, black is substituted.

3. Type of Unit Symbols

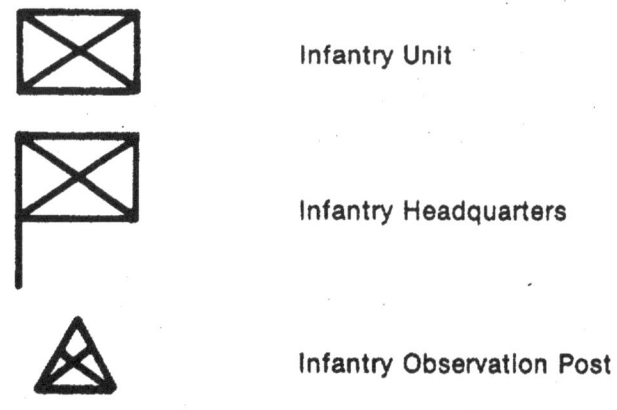

A-2

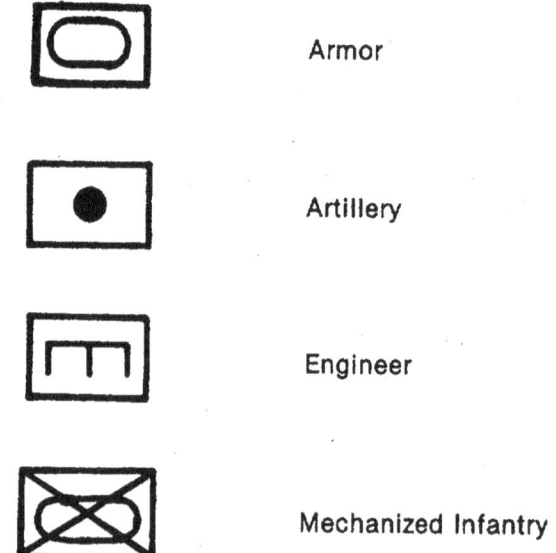

Armor

Artillery

Engineer

Mechanized Infantry

4. Sizes. Sizes of units are indicated by the following symbols.

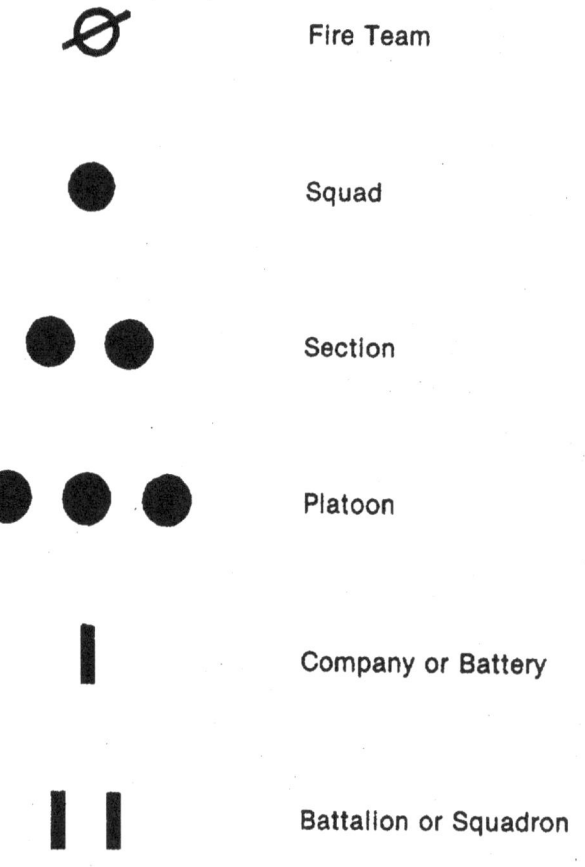

Fire Team

Squad

Section

Platoon

Company or Battery

Battalion or Squadron

5. Combining Basic Symbols

a. The arrangements of type unit symbols, size symbols, letters, and numbers to show specific troop units and observation posts are shown by the following diagram.

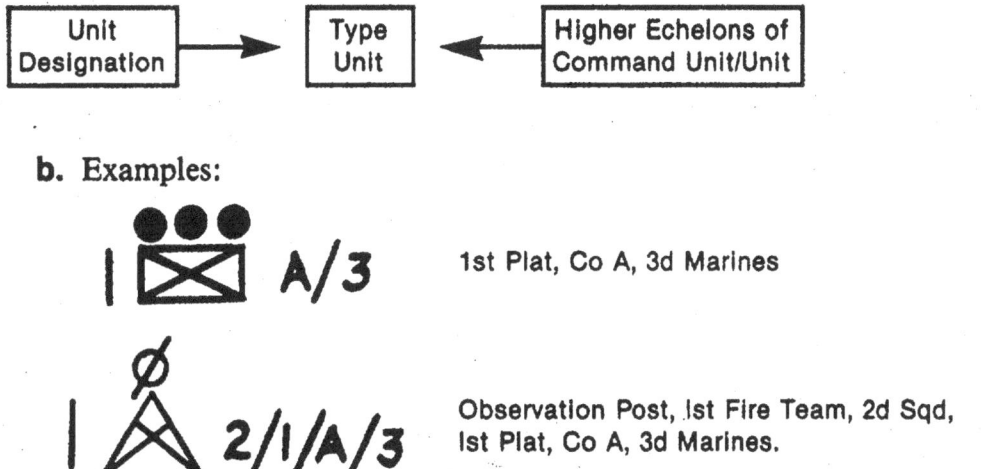

b. Examples:

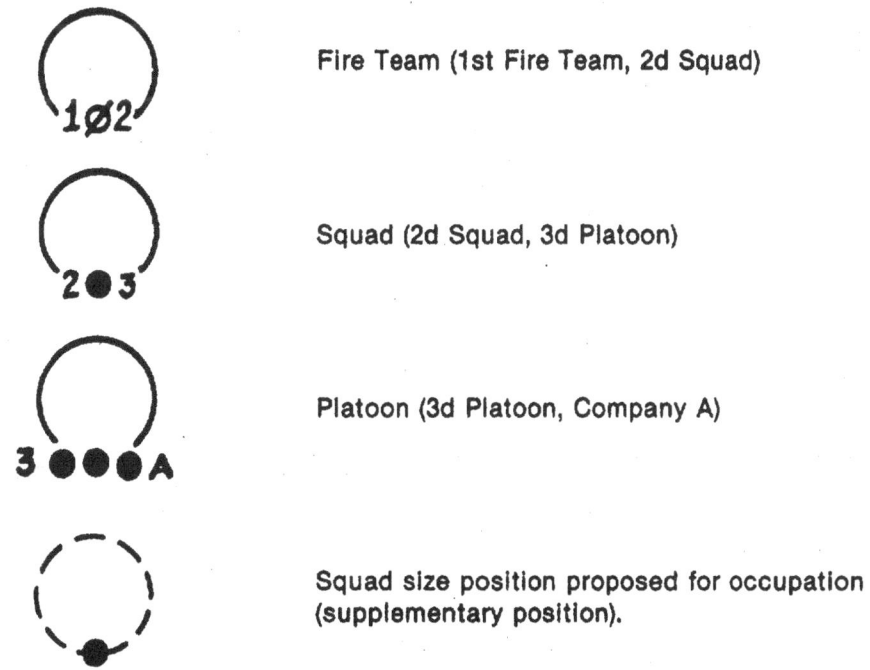

1st Plat, Co A, 3d Marines

Observation Post, 1st Fire Team, 2d Sqd, 1st Plat, Co A, 3d Marines.

6. Positions.
Positions of small infantry units are shown in this manner.

Fire Team (1st Fire Team, 2d Squad)

Squad (2d Squad, 3d Platoon)

Platoon (3d Platoon, Company A)

Squad size position proposed for occupation (supplementary position).

7. Weapons Symbols.
Following are some examples of weapons symbols:

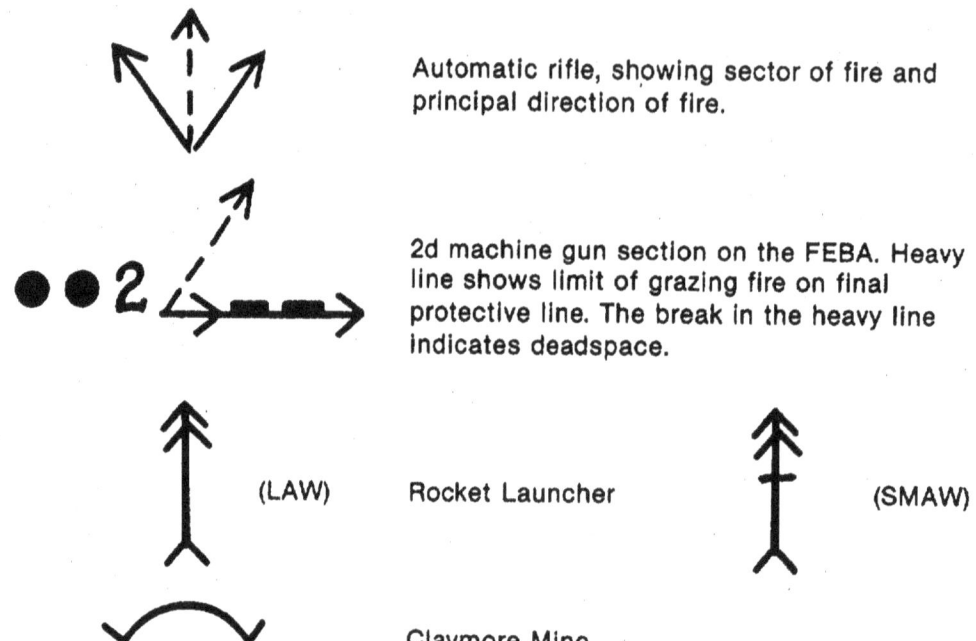

Automatic rifle, showing sector of fire and principal direction of fire.

2d machine gun section on the FEBA. Heavy line shows limit of grazing fire on final protective line. The break in the heavy line indicates deadspace.

(LAW)   Rocket Launcher   (SMAW)

Claymore Mine

8. Obstacle Symbols.
Following are some examples of obstacle symbols:

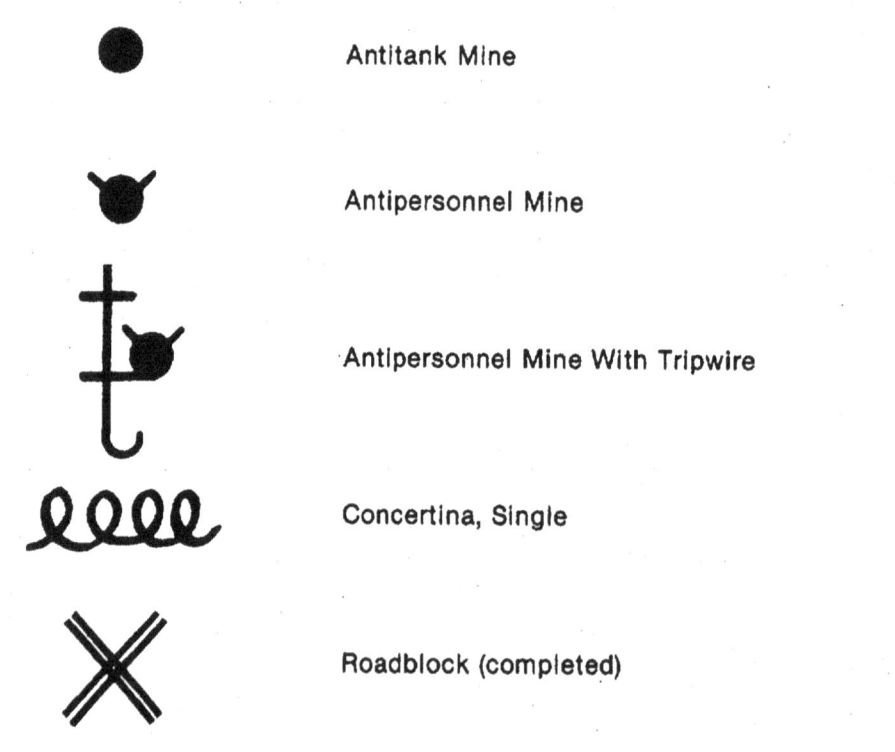

Antitank Mine

Antipersonnel Mine

Antipersonnel Mine With Tripwire

Concertina, Single

Roadblock (completed)

9. Miscellaneous Symbols. Following are some miscellaneous symbols:

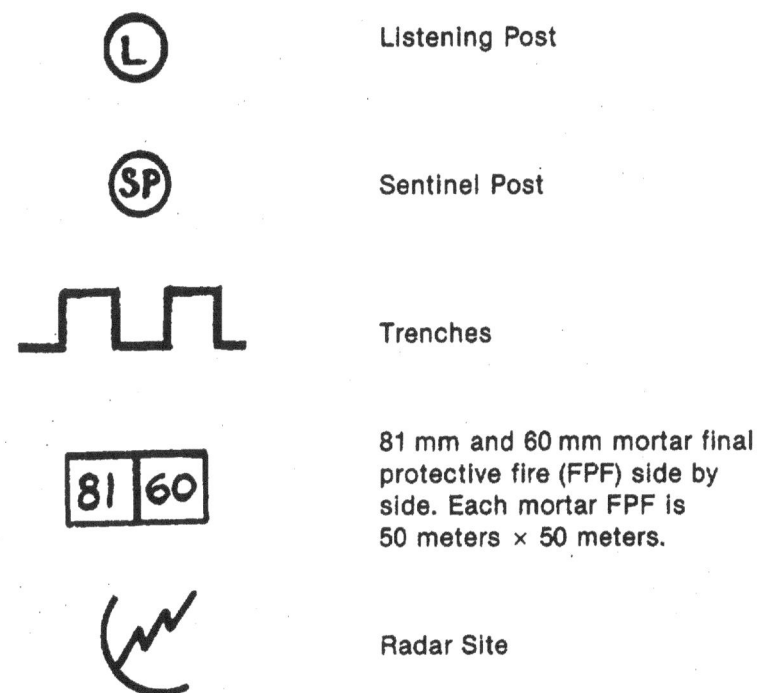

10. Enemy Positions. Enemy positions are shown by blue or black double lines if red is not available. Examples are as follows:

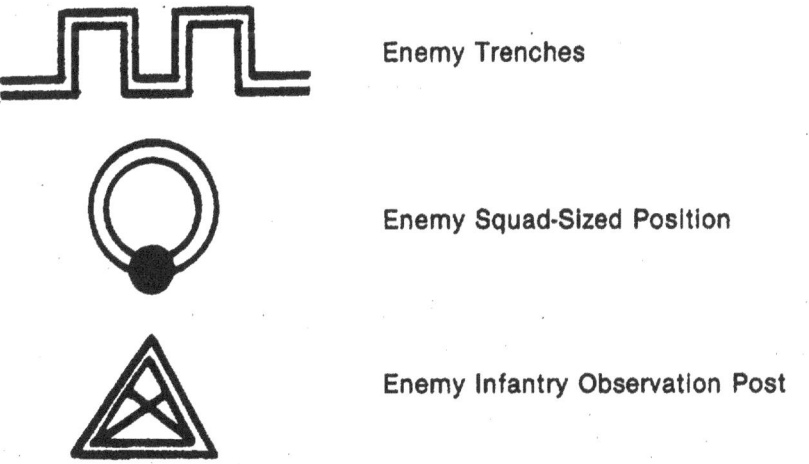

Appendix B

Method of Challenging by Sentries

1. Do not allow unidentified personnel to closely approach your position. Halt and identify them before they are close enough to be a danger to you. This precaution is important at night and during other periods of poor visibility.

2. The following definitions have been approved for Marine Corps use:

    a. Challenge. Any process carried out by one unit or person in order to determine the friendly or hostile identity of another.

    b. Reply. An answer to a challenge.

    c. Password. A secret word or distinctive sound given by a sentry.

    d. Countersign. A secret reply given in response to a sentry who has presented a password; for example, NUTS (password), WINE (countersign).

3. An example of how to use a challenge and reply, password and countersign is shown in Figure B-1.

ACTION BY SENTRY	ACTION BY PERSON OR GROUP CHALLENGED
1. HALT! WHO IS (OR GOES) THERE? (challenge)	1. Halts and gives any reply which indicates the person or group is authorized to pass; e.g., **FRIEND, ALLY, FIRE TEAM RETURNING FROM PATROL**, and so forth. (reply)
2. ADVANCE (ONE) AND BE RECOGNIZED.	2. Person (or group leader) advances without replying.
3. HALT! (sentry halts person when he is close enough to be recognized, or to give an unrecognized person the password). If the person is recognized by the sentry, he is allowed to pass. If not recognized, the sentry gives the password.	3. Person halts until recognized by sentry.
4. Password is given in a low tone.	4. Countersign is given in a low tone.
5. ADVANCE ANOTHER ONE (or REMAINDER) AND BE RECOGNIZED. Sentry calls forward remainder, singly or as a group, as the situation or his orders demand.	5. Next member (or remainder) of group advances at order of sentry to be recognized. Group leader, or person designated by leader, must remain with sentry to assist in identifying the remainder of the group.

Figure B-1. Procedure for Issuing Challenge and Reply, Password, and Countersign.

Appendix C

Troop Leading Procedures

1. General. The troop leading procedures listed below are aids in preparing for and executing assigned missions. They assist squad and fire team leaders in making the best use of time, facilities, and personnel. All the steps should be considered, but depending upon the mission and time available, the degree of consideration for each will vary.

2. Steps of Troop Leading Procedures (BAMCIS)

a. Begin Planning. When an order is received, the squad leader considers the time available to him. In so doing, he uses a planning sequence called **reverse planning,** meaning that he starts with the last action for which a time is specified (e.g., an attack) and works backward to the issuing of his order. This helps ensure that enough time is allowed for the completion of all necessary actions. During this stage, he also analyzes the terrain and the friendly and enemy situation. From his analysis, he formulates a preliminary plan of action to accomplish the mission. This plan is only tentative and will often be changed.

b. Arrange for Reconnaissance and Coordination. The squad leader selects a route and prepares a schedule for reconnaissance and coordination with adjacent and supporting units. Normally, he takes his fire team leaders and the leaders of any attached crew-served weapons teams with him on his reconnaissance.

c. Make Reconnaissance. On his reconnaissance, the squad leader completes his estimate of the situation. Prearranged meetings with adjacent squads and supporting units are held as scheduled. He notes how the terrain affects his preliminary plan and adopts, alters, or rejects it as necessary. While on his reconnaissance, he selects a vantage point from which to orient his fire team leaders.

d. Complete Plan. Upon his return from the reconnaissance, the squad leader completes his plan of action. He then prepares notes to be used in issuing his order.

e. Issue Order. If possible, the squad leader issues his order to the same personnel he took with him on his reconnaissance from the vantage point he had selected earlier. If this is not possible, the team leaders are oriented from maps, sketches, or an improvised terrain model. He issues his order using the five-paragraph order sequence and includes everything his fire team and attached weapons leaders need to know.

f. Supervise Activities. The squad leader continuously supervises his unit to ensure that his order is carried out as intended.

Appendix D
Estimate of the Situation

1. General. The estimate of the situation is a problem-solving process. It is a method of selecting the course of action which offers the greatest possibility of success. At the squad level, the estimate is a continuous, rapid mental process and should be followed no matter how quickly a decision must be made.

2. METT-T. The squad leader analyzes the courses of action and considers the advantages and disadvantages of each by using the following **yardstick**:

- Mission (M)
- Enemy (E)
- Terrain and weather (T)
- Troops and support available (T)
- Time available (T)

3. Estimate of the Situation Process. The squad leader selects the best course of action by applying the factors of METT-T to each possible course of action. This forms the basis of the squad leader's decision. An explanation of the factors represented in METT-T follows.

a. Mission. The mission is a clear, concise, and simple statement of the task to be performed. It must be carefully examined and thoroughly understood. It is the basis for all actions of the squad until it is accomplished.

b. Enemy. Information concerning the enemy comes from many sources. The most reliable information is obtained by personal reconnaissance and, time permitting, no decision should be made without a reconnaissance. The squad leader's aim is to find out the enemy's location, strength, composition, type of weapons, disposition, tactical methods, and recent actions.

c. Terrain and Weather. The terrain and weather affect all plans and actions. They must be studied from both the friendly and enemy viewpoints. The squad leader's plan of action must take full advantage of the terrain. The weather, both present and predicted, will affect visibility, movement, and fire support. The military aspects of terrain (often referred to as KOCOA) are as follows:

(1) Key Terrain. Key terrain is any feature or area which gives a marked advantage to the force controlling it. Generally, this advantage lies in terrain which affords good observation and fields of fire.

(2) Observation and Fields of Fire. Observation is the ability to view enemy locations or avenues of approach in order to gain information about or direct accurate fire onto the enemy. Fields of fire are the areas that a weapon or group of weapons can cover and are essential to the effective employment of direct fire weapons. Observation and fields of fire should be considered both from friendly and enemy points of view.

(3) Cover and Concealment. Cover is protection from enemy fire. Concealment is the hiding or disguising of a unit and its activities from enemy observation. Terrain features that offer cover also provide concealment. The greater the irregularity of the terrain, the more concealment is offered from ground observation.

(4) Obstacles. Obstacles are natural or artificial features which stop, delay, or restrict military movement. They may either help or hinder a unit, depending upon their location and nature. For example, a deep creek located across the direction of movement will slow an attacker, while the same type of creek on the flank of an attacker affords a measure of security. In general, obstacles perpendicular to the direction of movement favor the defender, while those parallel to the direction of movement may give the attacker an advantage by protecting his flanks and providing him with covered avenues of approach.

(5) Avenues of Approach. An avenue of approach is terrain which provides a force a route of movement. It should also provide ease of movement, cover and concealment, favorable observation, fields of fire, and adequate maneuver room.

d. Troops and Fire Support Available. The squad leader considers his unit's strength and abilities against that of the enemy. He should know what assistance he has available from supporting weapons (machine guns, rocket launchers, mortars, tanks, artillery, naval gunfire, and aircraft).

e. Time Available. The efficient use of time is always critical to success. The squad leader must determine how much time he has to plan and execute the required tasks. Time must not be wasted; on the other hand, the squad leader must not allow tasks to be rushed to the extent that they are done incompletely or not at all. When time is short, tasks must still be accomplished completely.

Appendix E

Squad Five-Paragraph Order

The five-paragraph order is derived from the operation order and is structured to meet the needs of the small-unit leader. It is similar to the operation order in that it includes the situation, mission, concept of operation, orders to subordinates, and measures required to ensure coordination of administrative, logistic, command, and communication matters. The five-paragraph order is structured for oral presentation while the operation order is structured to be writtten. The five-paragraph order is used at company level and below. The squad leader issues his order orally.

Five-Paragraph Order Format (SMEAC)

1. Situation

   a. <u>Enemy Forces</u>. Consists of the composition, disposition, location, movement, capabilities, and recent activities of enemy forces.

   b. <u>Friendly Forces</u>. A statement of the mission of the next higher unit, location and mission of adjacent units, and mission of nonorganic supporting units which may affect the actions of the unit.

   c. <u>Attachments and Detachments</u>. Units attached to or detached from the squad by higher headquarters, including the effective time of attachment or detachment.

2. Mission

A clear, concise statement of the task which the squad must accomplish.

3. Execution

   a. <u>Concept of Operations</u>. The concept of operation is the squad leader's brief summary of the tactical plan the squad is to execute.

b. <u>Subordinate Tasks (Missions)</u>. In each succeeding paragraph, missions are assigned to each fire team and any attached units.

c. <u>Reserve</u>. This paragraph identifies the unit assigned the reserve mission and the tasks assigned to the reserve. Normally, a reserve is not designated below the company level. If no reserve is designated, this paragraph is omitted.

d. <u>Coordinating Instructions</u>. In the last paragraph, instructions that apply to two or more subordinate units are given.

4. <u>Administration and Logistics</u>

This paragraph contains information or instructions pertaining to rations and ammunition; location of the distribution point, corpsman, and aid station; the handling of prisoners of war; and other administrative and supply matters.

5. <u>Command and Signal</u>

a. Special instructions on communications, including prearranged signals, password and countersign, radio call signs and frequencies, emergency signals, radio procedures, pyrotechnics, and restrictions on the use of communications.

b. Location of the platoon commander.

c. Location of the platoon sergeant.

d. Location of the squad leader.

Appendix F

Reporting Information

1. General. Information must be reported quickly, accurately, and as completely as possible. The acronym **SALUTE** provides a simple method for remembering how and what to report about the enemy.

Size
Activity
Location
Unit*   * (Enemy unit may be derived from unit
Time   markings, uniform, or through prisoner-
Equipment   of-war interrogation.)

An example of such a report is "Seven enemy soldiers, traveling SW, crossed road junction on Black Ridge at 211300 August. They were wearing green uniforms and carrying one machine gun and one rocket launcher."

2. Shelling Reports (SHELREP). The squad should report enemy artillery and mortar fire, and aircraft bombings using a SHELREP. The following format is suitable for either a written or oral report:

Alfa	Observer's call sign.
Bravo	Observer's location.
Charlie	Azimuth to enemy gun.
Delta	Time shelling started.
Echo	Time shelling stopped.
Foxtrot	Coordinates of area shelled, if a map is available.
Golf	Number and types of weapons fired.
Hotel	Nature of fire: destruction, harassing, or registration.
India	Number and type of shells.
Juliet	Flash-bang time in seconds.
Kilo	Damage (usually in code).

3. Estimating Range by the Flash-Bang Method. Sound travels about 330 meters (1,100 feet) per second. When the observer sees the flash or smoke of a weapon, or the dust it raises, he starts counting seconds (one thousand one,

one thousand two, and so forth). He stops counting when he hears the report of the weapon. If he stops on the count of *one thousand three,* for example, the range from the observer to the gun is three times 330 meters per second or 990 meters (3,300 feet). Marines should practice timing their count with the second hand of a watch to develop the correct speed.

4. Crater Analysis. (See fig. F-1.) If an observer is unable to determine the location of a gun by direct observation, he may be able to determine the line of flight of the projectile by examining the crater. Also, the type and caliber of weapon may be determined from the identification of shell or fuze fragments and tail fins found in the crater. Information on the line of flight and type and caliber of the projectile is passed to the platoon commander. Although not difficult to learn, crater analysis does require some training. Detailed procedures for conducting crater analysis is found in FMFM 6-8, *Supporting Arms Observer, Spotter, and Controller.*

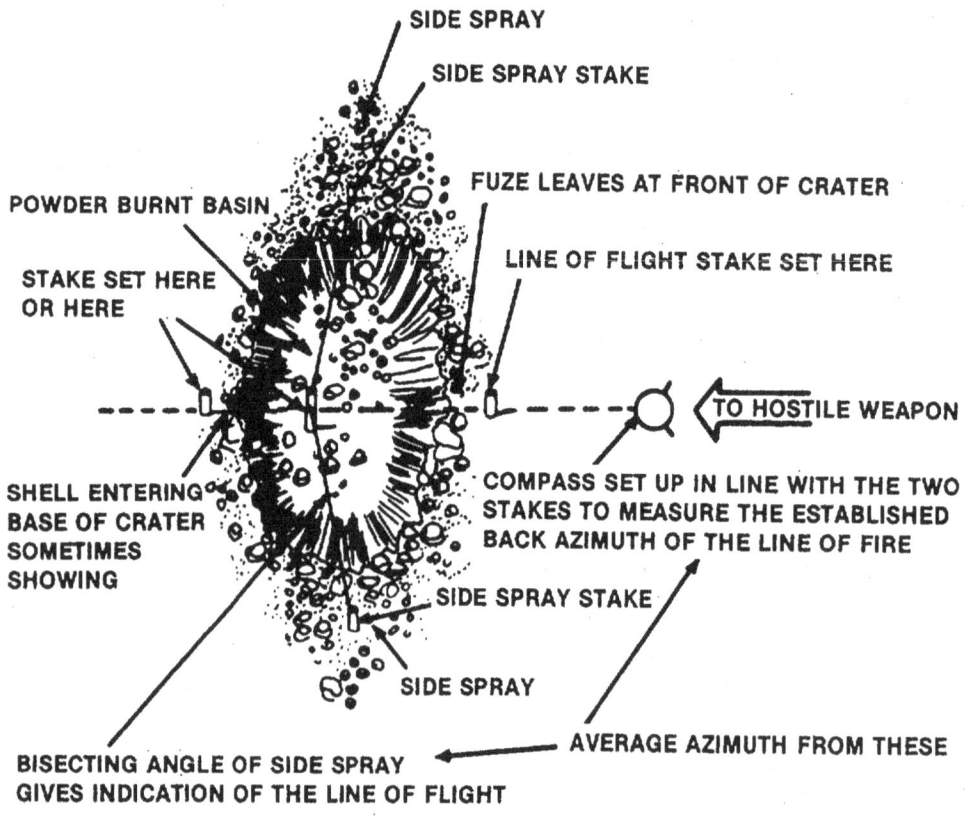

Figure F-1. Example of Crater Analysis.

Appendix G

Handling Prisoners of War

1. General. Prisoners of war (POWs) are one of the most valuable sources of information for intelligence purposes. They should be delivered to the platoon commander as quickly as possible.

2. The Five Ss. The five Ss of POW handling are:

a. Search. Prisoners are thoroughly searched for weapons and documents as soon as they have been captured. Weapons and documents should be tagged and immediately sent to the platoon commander.

b. Segregate. Prisoners are segregated into groups: officers, NCOs, privates, deserters, civilians, and females. This prevents leaders from organizing escapes and issuing orders to subordinates.

c. Silence. Silence is essential. Prisoners must not be allowed to talk to one another.

d. Speed. Speed is required in getting prisoners to the platoon commander. Timely information secured from prisoners is essential.

e. Safeguard. Prisoners are safeguarded as they are moved. They are restrained, but not abused. They are not given cigarettes, food, or water until authorized by assigned interrogators.

Appendix H

Cover, Concealment, and Camouflage

1. Each Marine must use terrain to give himself cover and concealment. He must supplement natural cover and concealment with camouflage.

2. Cover is protection from the fire of enemy weapons. It may be natural or manmade. (See fig. H-1.)

Figure H-1. Cover.

a. Natural cover includes logs, trees, stumps, ravines, hollows, reverse slopes, and so forth. Manmade cover includes fighting holes, trenches, walls, rubble, abandoned equipment, and craters. Even the smallest depression or fold in the ground gives some cover. Marines must look for and use every bit of cover the terrain offers.

b. When the enemy approaches a fighting position and brings it under direct and indirect fire, there must be cover to protect the troops. Natural cover is best as it is the most difficult for the enemy to spot.

c. Marines dig fighting holes to increase the protection afforded by natural cover against enemy direct and indirect fires. The type and extent of preparation will depend on the mission and the length of stay.

d. First, the Marine prepares a simple prone shelter. Then, as time allows, he prepares a more fully developed position, up to a complete fighting hole with overhead cover and trenches connecting it with other positions.

e. When moving, Marines use routes which put cover between them and places where the enemy is known or thought to be. They use ravines, gullies, hills, wooded areas, and other natural cover to keep the enemy from seeing and shooting at them. They avoid open fields. Units avoid skylining on hills and ridges. In a desert, rock formations and depressions are cover. (See fig. H-2.)

Figure H-2. Covered Route

3. Concealment is anything that hides a Marine, his position, unit, or equipment from enemy observation. Small-unit leaders must enforce light and noise discipline, control movement, and supervise the use of camouflage. Well-hidden fighting holes help conceal a unit's location from the enemy. The best way to use natural concealment is to refrain from disturbing it when moving into an area. Darkness alone does not hide a unit from an enemy who has night vision and other detection devices. (See fig. H-3.)

Figure H-3. Concealed Position.

4. Camouflage makes use of natural and manmade material. Used well, it reduces the chance of detection by the enemy. If camouflage material is needed, it should be brought from outside the fighting position. If used well, branches, bushes, leaves, and grass provide the best camouflage. Foliage used as camouflage must blend with that of the surrounding area. An open, exposed position can be concealed from enemy observation by using the right materials and procedures.

5. Some things the enemy will look for in trying to find friendly positions are listed below.

a. Movement draws attention. An observer will catch movement in his field of view. Moves, such as arm-and-hand signals, can be seen by the naked eye at long ranges. A comparison of aerial photos taken of the same area at different times can reveal movement of troops and vehicles and will help the enemy find targets. (See fig. H-4.)

b. Shadows draw attention. Camouflage should be used to break up shadows of fighting positions and equipment. Shaded areas offer concealment. This is particularly true of shadows of buildings in cities. (See fig. H-5.)

Figure H-4. Movement.

Figure H-5. Shadows.

c. Fighting positions should not be where the enemy expects to find them. They should be on the side of a hill, away from road junctions or lone buildings, and in covered and concealed locations. (See fig. H-6.)

Figure H-6. Fighting Positions.

d. Shape is the outline of something. The shape of the helmet is easily recognized, as is the undisguised shape of a man's body. Both camouflage and concealment should be used to make familiar shapes blend with their surroundings. (See fig. H-7.)

Figure H-7. Shape.

e. Shine may be a light source such as a cigarette glowing in the dark, or reflected light from smooth, polished surfaces such as a worn metal surface, a windshield, binoculars, eyeglasses, a watch crystal, or exposed skin not toned down with face paint. The use of lights or the reflection of light may help the enemy detect friendly positions. Equipment that shines should be concealed, or covered with mud or paint. (See fig. H-8.)

Figure H-8. Shine.

f. Contrasting colors are more easily detected; for example, against the dark green of jungle foliage, white skin shows up better than does black. Camouflage should match the surrounding area, rather than offer a contrast. Bright colors should not be used in camouflage.

g. Dispersion is the distance between men, vehicles, and equipment. If a squad is not dispersed, it is easier to detect and easier to hit. Distances between men, teams, and squads must be prescribed and enforced.

6. How to Camouflage

a. Before camouflaging, Marines study the terrain and vegetation of the area they are in and the area to which they are going. Grass, leaves, brush, and other natural materials must be arranged to conform to the area. Tree branches stuck into the ground in an open field will not fool anyone. Vegetation changes from area to area. As units move from one area to another, camouflage must be changed to blend with the vegetation.

b. Marines should camouflage or hide dirt from fighting holes and heads. If necessary, they take it away from the positions and camouflage it. (See fig. H-9.)

Figure H-9. Hiding Dirt.

c. Marines should use only that material which is needed. Too much camouflage (natural or manmade) may call attention to a position as easily as will too little. Camouflage materials should be gathered from a wide area. An area stripped of all its foliage will draw attention. (See fig. H-10.)

d. Dirt used in parapets and overhead cover must be camouflaged. If the fighting holes have no overhead cover, the bottom of the holes must also

Figure H-10. Natural Positions.

be camouflaged to prevent detection from the air. When possible, open areas should be covered by fire rather than physically occupied since it is hard to conceal a position in the open.

e. Men must continue to camouflage their positions as they prepare them. Work on a defensive position in daylight depends on the enemy air threat and whether or not the enemy can see the position. When the enemy has air superiority, work may be possible only at night. Shiny or light-colored objects which attract attention from the air must not be left lying about. Mirrors, food containers, towels, etc., must all be hidden. Shirts must not be removed, as the exposed skin stands out and increases the chance of being seen. Fires must not be used where there is a chance that the smoke or flame will be seen by the enemy. Trails and other evidence of movement must be hidden.

f. After camouflage is complete, the fighting position should be inspected from the enemy's point of view. Camouflage should be checked often to see that it stays natural looking and conceals the position. If it looks like a camouflaged position to the Marine inspecting, it is almost certain that it will look like a camouflaged position to the enemy.

g. Helmets must be covered with the issue helmet cover or one made of cloth or burlap colored to blend with the terrain. The cover should fit loosely. Foliage should stick over the edges. This should not be overdone. If there is no material for helmet covers, the surface of helmets can be disguised and dulled with irregular patterns of paint or mud. Camouflage bands, string, burlap strips, or rubber bands can be used to hold the foliage in place. (See fig. H-11.) Uniforms must blend with the terrain. Badly faded equipment may be hard to hide. Units should turn in badly faded equipment or use mud, a camouflage stick, paint, and so forth, to color it until it can be exchanged.

h. When operating in snow-covered terrain, Marines should wear overwhites and, where possible, they should color equipment white. If overwhites are not issued, sheets or other white cloth can be used for camouflage. (See fig. H-12.)

i. Exposed skin reflects light and draws the enemy's attention. Even very dark skin, because of its natural oil, will reflect light. Camouflage face paint sticks are issued in three standard two-tone sticks. (See fig. H-13.)

(1) When applying camouflage face paint, men work with one another. They check each other. They apply a two-color combination in an irregular pattern. Shine areas (forehead, cheekbones, nose, ears, and chin) are painted with a dark color. Shadow areas (around the eyes, under the nose, and under the chin) are painted with a light color. Exposed skin on the back of the neck, ears, arms, and hands should be painted.

(2) When face paint sticks are not issued, mud or charcoal can be used to tone down exposed skin.

Figure H-11. Foliage.

Figure H-12. Blending.

Figure H-13 Camouflage Sticks.

Appendix I

Glossary

Acronyms

AAV .... assault amphibious vehicle
ADDRAC .... alert, direction, description, range, assignment, control
APC .... armored personnel carrier

CCW .... Certain Conventional Weapons Which May be Deemed to be
Excessively Injurious or to Have Indiscriminate Effects
(otherwise known as the CCW Convention)
CP .... command post

FASCAM .... family of scatterable mines
FCL .... final coordination line
FEBA .... forward edge of the battle area
FPF .... final protective fire
FPL .... final protective line

HE .... high explosive
HEDP .... high explosive dual purpose

IA .... immediate action

KOCOA .... key terrain, observation, cover and concealment,
obstacles, and avenues of approach

LAW .... light assault weapon
LP .... listening post

METT-T .... mission, enemy, terrain and weather, troops and
support available, and time available
MOUT .... military operations on urbanized terrain

NCO	noncommissioned officer
NSDAPL	non-self-destructing/deactivating antipersonnel landmines
NSDATL	non-self-destructing/deactivating antitank landmines
OP	observation post
PDF	principal direction of fire
PLD	probable line of deployment
POW	prisoner of war
RPM	rounds per minute
SAW	squad automatic weapon
SDAPL	self-destructing/deactivating antipersonnel landmines
SDATL	self-destructing/deactivating antitank landmines
SHELREP	shelling report
SMAW	shoulder-launched, multipurpose assault weapon
SOP	standing operating procedure
WP	white phosphorus

Appendix J

References

1. Fleet Marine Force Manuals and Reference Publications

FMFM 0-8	Basic Marksmanship (under development)
FMFM 1-2	Marine Troop Leader's Guide
FMFM 6-4	Marine Rifle Company/Platoon
FMFM 6-7	Scouting and Patrolling for Infantry Units
FMFRP 0-1A	How to Conduct Training
FMFRP 0-1B	Marine Physical Readiness Training for Combat
FMFRP 2-72	Multi-Service Procedures for the Application of Firepower (J-FIRE)
FMFRP 7-23	Small-Unit Leader's Guide to Cold Weather Operations

2. U.S. Army Field Manuals

FM 21-15	Care and Use of Individual Clothing and Equipment
FM 21-18	Foot Marches
FM 21-26	Map Reading and Land Navigation
FM 21-60	Visual Signals
FM 21-75	Combat Skills of the Soldier
FM 21-76	Survival
FM 23-14	Squad Automatic Weapon (SAW), M-249
FM 23-23	Antipersonnel Mine, M-18A1 and M-18 (Claymore)
FM 23-30	Grenades and Pyrotechnic Signals
FM 23-31	40mm Grenade Launchers, M-203 and M-79
FM 23-33	66-mm HEAT Rocket, M-72A1 and M-72A2 (Light Antitank Weapon)
FM 101-5-1	Operational Terms and Symbols

Index

                                        Paragraph   Page

A

Entry	Paragraph	Page
Ambushes:		
Classifications	8506b	8-37
Execution	8506e	8-40
Miscellaneous techniques	8506f	8-41
Mission	8502d	8-34
Purpose of	8506a	8-36
Success factors	8506d	8-38
Types	8506c	8-37 — 8-38
Ammunition:		
M-16 rifle	1004b(1)	1-3
M-203 grenade launcher	1004b(3)	1-3 — 1-4
Squad automatic weapon	1004b(2)	1-3
Amphibious operations:		
Assault	6006	6-8
Debarkation	6004	6-3 — 6-7
Debarkation procedure from transports	6004e	6-6
Duties aboard ship	6003	6-1 — 6-2
Equipment preparations, individual	6004c	6-5 — 6-6
Lashing and lowering equipment	6004f	6-6
Movement from assembly area to debarkation station	6004d	6-6
Movement from ship-to-shore	6005	6-7 — 6-8
Preembarkation	6002	6-1
Purpose	6001	6-1
Application of fire	2601	2-26
Arm-and-hand signals	3204	3-35 — 3-51
Assault amphibious vehicle:		
Organization and composition	6004a(2)	6-3
Preparation, boat team	6004g	6-7
Assistant automatic riflemen:		
Duties	1005e	1-5
Fire delivery:		
In the attack	2603b(2)(a)	2-32
In the defense	2603c(1)	2-33
Weapons	1004a(4)	1-1

Assistant boat team commander, duties	6004b(2)	6-4
Automatic riflemen:		
Duties	1005d	1-5
Fire delivery:		
In the attack	2603b(2)(b)	2-32
In the defense	2603c(1)	2-33
Target assignment	2507b(1)	2-22
Weapons	1004a(3)	1-1

B

Battle position	5103c, Fig. 5-2	5-4, 5-5
Beaten zone	2304	2-8
Boat paddle handler, duties	6004b(5)	6-5
Boat team:		
Duties of personnel:		
Assistant commander	6004b(2)	6-4
Boat paddle handler	6004b(5)	6-5
Commander	6004b(1)	6-3 — 6-4
Loaders	6004b(3)	6-4
Net	6004b(4)	6-4
Organization	6004a	6-3

C

Camouflage	App. H	H-3 — H-12
Captured documents	8603	8-43
Challenging, procedure	App. B, Fig. B-1	B-1 — B-2, B-2
Chemical defense:		
Decontamination	9502d	9-41 — 9-42
Detection of chemical agents	9502b	9-40 — 9-41
Protection	9502c	9-41
Types of chemical agents	9502a	9-38 — 9-39
Classes of fire with respect to the target	2305a	2-8 — 2-10
Clearing fields of fire	5204d	5-21 — 5-23
Colors, in military symbols	App. A	A-1

Combat formations	3101	3-1
Changing	3103	3-12–3-34
Column to echelon left	Fig. 3-12e	3-17
Column to echelon right	Fig. 3-12d	3-16
Column to skirmishers left	Fig. 3-12c	3-15
Column to skirmishers right	Fig. 3-12b	3-14
Column to wedge	Fig. 3-12a	3-13
Echelon left to skirmishers left	Fig. 3-12u	3-33
Echelon left to wedge	Fig. 3-12s	3-31
Echelon right to column	Fig. 3-12q	3-29
Echelon right to skirmishers right	Fig. 3-12t	3-32
Echelon right to wedge	Fig. 3-12r	3-30
Fire team	3102a	3-1
Skirmishers left to column	Fig. 3-12l	3-24
Skirmishers left to echelon left	Fig. 3-12p	3-28
Skirmishers left to wedge	Fig. 3-12n	3-26
Skirmishers right to column	Fig. 3-12k	3-23
Skirmishers right to echelon right	Fig. 3-12o	3-27
Skirmishers right to wedge	Fig. 3-12m	3-25
Wedge to column	Fig. 3-12f	3-18
Wedge to echelon left	Fig. 3-12j	3-22
Wedge to echelon right	Fig. 3-12i	3-21
Wedge to skirmishers left	Fig. 3-12h	3-20
Wedge to skirmishers right	Fig. 3-12g	3-19
Fire Team:		
Column	3102a(1), Fig. 3-1	3-1, 3-2
Echelon	Fig. 3-4	3-5
Echelon right (left)	3102a(4)	3-4
Skirmishers right (left)	3102a(3), Fig. 3-3	3-3, 3-4
Wedge	3102a(2), Fig. 3-2	3-2, 3-3
Squad:		
Column	3102b(1), Fig. 3-5	3-5–3-7, 3-6
Echelon	3102b(5)	3-8
Echelon left	Fig. 3-11	3-11
Echelon right	Fig. 3-10	3-10
Line	3102b(4), Fig. 3-8, Fig. 3-9	3-8, 3-9, 3-9

Combat formations—Continued
- Vee .... 3102b(3), 3-7—3-8,
  - Fig. 3-7   3-8
- Wedge .... 3102b(2), 3-7,
  - Fig. 3-6   3-7

Combat patrols:
- Ambush patrols. *See* Ambushes.
- Contact patrols 8501   8-34
- Equipment 8504   8-35
- Security patrols. *See* Security patrols.
- Task organization 8503   8-34
- Types and missions 8502   8-34

Combat signals 3201   3-35
- Arm-and-hand:
  - Combat formations .... 3204a 3-35—3-47
  - Helicopter operations .... 3204b 3-48—3-51
- Special 3203   3-35
- Whistle 3202   3-35

Composition of rifle squad 1003, 1-1,
  Fig. 1-1   1-2
Concealment .... App. H H-3
Cone of fire 2303   2-7—2-8
Contact patrols 8501   8-34
- Actions at the objective 8505c 8-36
- Mission 8502b 8-34
- Organization and equipment 8505b 8-35
Cover .... App. H H-1—H-2
Crater analysis .... App. F F-2

D

Defensive combat:
- Against mechanized attack 5207   5-32
- Camouflage measures .... 5204h 5-27—5-28
- Clearing fields of fire .... 5204d, 5-21—5-23,
  Fig. 5-11 5-22
- Conduct 5206   5-30—5-31
- Defense order, squad 5205   5-28—5-30
- Definitions 5103   5-2—5-7
- Fire plan, rifle squad 5107b 5-14
- Fire plan sketch, squad 5107f, 5-17,
  Fig. 5-10 5-17

Defensive combat—Continued
  Fire team:
    Assistant automatic riflemen . . . . 5106b(5)   5-11
    Automatic riflemen . . . . 5106b(3)   5-11
    Fire plan . . . . 5106b,   5-10—5-13,
                                                         Fig. 5-5   5-10
    Fire plan sketch . . . . 5106b(9),   5-13,
                                                         Fig. 5-6   5-13
    Fire plan sketch symbols . . . . Fig. 5-7   5-14
    Fire team leader, position of . . . . 5106b(6)   5-11
    M-203 employment . . . . 5106b(7)   5-12
    Organization of the ground 5106a   5-9
    Riflemen . . . . 5106b(4)   5-11
    Sector of fire . . . . Fig. 5-4,   5-9,
                                                         5106b(8)   5-12
  Fundamentals 5104   5-7—5-8
  Ground organization 5204   5-20—5-28
    Fighting holes . . . . 5204e   5-23—5-26
  Local security for platoons and companies 5209   5-33
  Mission 5102   5-1
  Missions of the squad:
    Frontline squad 5105a   5-8
    Squad as a security element 5105c   5-9
    Squad as part of the reserves . . . . 5105b   5-8—5-9
  Movement to supplementary fighting positions . . 5208   5-32
  Organization of the ground 5107a   5-14
  Plan of defense 5202   5-18—5-20
  Purpose 5101   5-1
  Security forces 5210   5-33—5-35
  Signals:
    Cease final protective fires 5202c   5-20
    Commence final protective fires 5202b   5-20
    Commence firing 5202a   5-19
  Squad fighting positions . . . . Fig. 5-8   5-15
  Squad security 5203   5-20
  Supplementary and alternate fighting positions . . 5204g   5-27
  Troop leading procedures 5201   5-18
Demolitions:
  Accessories . . . . 9402e   9-32—9-33
  Explosives 9402c   9-32
  Prepared charges 9402d   9-32
  Training 9402a   9-31

Duties	1005	1-4
Assistant automatic rifleman	1005e	1-5
Automatic rifleman	1005d	1-5
Fire team leader/grenadier	1005c	1-5
Rifleman	1005f	1-5
Squad leader	1005b	1-5

E

Economy of force patrols	8502c	8-34
Enemy positions, symbols	App. A	A-5
Estimate of the situation	App. D	D-1 – D-3
Explosives	9402c	9-32
Explosives, characteristics	9402b	9-32

F

Fighting holes:		
Digging	5204e	5-23 – 5-26
One-man	5204e(1), Fig. 5-12, Fig. 5-13	5-23 – 5-26, 5-24, 5-24
One-man, protection against tanks	Fig. 5-14	5-25
Two-man	5204e(2), Fig. 5-15	5-26, 5-26
Fighting position	5103b	5-3 – 5-4
Alternate	5103b(2)	5-3
Primary	5103b(1)	5-3
Supplementary	5103b(3)	5-3 – 5-4
Fire:		
Classes of	2305	2-8 – 2-11
Control and discipline	2606	2-35
Delivery	2603	2-30 – 2-33
Effect of	2306	2-11
Enfilade	2305a(3)	2-10
Flanking	2305a(2)	2-8
Frontal	2305a(1)	2-8
Grazing	2305b(1)	2-10
Overhead	2305b(3)	2-11
Plunging	2305b(2)	2-10
Preplanned	2604, Fig. 2-18	2-33, 2-34

Fire—Continued
   Rates 2605   2-34
      Average 2307a   2-11
      Rapid 2307c   2-12
      Sustained 2307b   2-12
   Reduced visibility 2604   2-33
   With respect to the ground .... 2305b,   2-10—2-11,
                                    Fig. 2-8   2-10
   With respect to the target .... Fig. 2-6,   2-9,
                                    Fig. 2-7   2-9

Fire commands:
   Alert 2503   2-18
   Delivery 2510   2-23—2-24
   Direction:
      Finger measurement .... 2504d   2-20—2-21
      Oral 2504a   2-19
      Reference points 2504c   2-20
      Tracer ammunition 2504b   2-19
   Elements 2502   2-17
   Fire control 2508   2-23
   Purpose and importance 2501   2-17
   Range 2506   2-21
   Sector of fire .... Fig 5-4   5-9
   Signals 2509   2-23
   Subsequent 2511   2-24—2-25
   Target assignment:
      Assignment to teams 2507a   2-21
      Automatic riflemen .... 2507b(1)   2-22
      Examples .... 2507c   2-22—2-23
      Fire team leader/grenadier .... 2507b(2)   2-22
   Target description 2505   2-21
   Topographical terms .... Fig. 2-11   2-18

Fire delivery:
   In the attack .... 2603b   2-31—2-33
      Assault fire .... 2603b(2)   2-31—2-33
      Assault fire positions .... Fig. 2-17   2-32
      Base of fire .... 2603b(1)   2-31
   Position requirements .... 2603a   2-30—2-31

Fire plan:
   Fire team .... 5106b   5-10—5-13
   Rifle squad 5107b   5-14

Fire plan—Continued
  Sketch:

Fire team	5106b(9), Fig. 5-6	5-13, 5-13
Squad	5107f, Fig. 5-9	5-17, 5-16
Fire support	2601b	2-26
Battalion	1006c	1-6
Company	1006b	1-6
Platoon	1006a	1-6

Fire team:
  Combat formations. *See* Combat formations, fire team.

Composition	1003	1-1

  Defensive combat. *See* Defensive combat, fire team.

Fire plan	5106b	5-10—5-13
Fire plan sketch	5106b(9)	5-13
Leader	1003	1-1
Offensive combat, movement from the line of departure to the assault position	4302	4-22

Fire team leader, fire delivery:

In the attack	2603b(2)(c)	2-32—2-33
In the defense	2603c(2)	2-33

Fire team leader/grenadier:

Duties	1005c	1-5
Target assignment	2507b(2)	2-22
Weapons	1004a(2)	1-1
Five-paragraph order	App. E	E-1—E-2
Fortified areas	9201	9-15—9-20

  Attacking:

Assault element	9201e	9-19—9-20
Base of fire, employment of	9201d	9-18
Rifle squad role	9201c	9-18
Tasks	9201b	9-16—9-18
Characteristics	9201a	9-15—9-16
Seizing an objective	9202	9-20
Assault of the emplacement	9202f	9-22
Movement to the assault position	9202e	9-22
Planning and coordination	9202b	9-20
Planning the attack	9202c	9-20—9-21
Preparation fires	9202d	9-21
Forward edge of the battle area	5103d	5-4

G

Guerrilla operations	9601	9-43
Attacking guerrilla houses	9608	9-49 — 9-50
Characteristics	9602	9-43
Control over civil populace	9605	9-47
Indoctrination for the squad	9603	9-43 — 9-44
Offensive action	9607	9-48 — 9-49
Patrol bases	9604	9-44 — 9-47
Patrol operations against	9606	9-48

H

Helicopterborne operations	7001	7-1
Concept of employment	7002	7-1
Conduct of the assault	7005	7-5 — 7-6
Initial ground action	7005b	7-6
Planning	7005a	7-5
Helicopter team	7003a	7-1
Helicopter wave	7003b	7-1
Landing point	7003e	7-2
Landing site	7003d	7-2
Landing zone	7003c	7-2
Training, heliteam	7004	7-2 — 7-5
Assistant leader's responsibility	7004c	7-3
Landing	7004g	7-5
Loading	7004e	7-4 — 7-5
Loading aboard ship	7004f	7-5
Loading procedure	7004d,	7-3
Organization	7004a	7-3
Pickup zone procedure	Fig. 7-2	7-4
Team leader's responsibility	7004b	7-3

I

Infantry-tank coordination. *See* Tank-infantry coordination.		
Information reporting	App. F	F-1
Crater analysis	App. F	F-2
Range estimation	App. F	F-1 — F-2
Shelling reports	App. F	F-1

L

Landmines	9401c	9-29 — 9-30
Leading troops, procedures	App. C	C-1 — C-2
Loaders, duties	6004b(3)	6-4

M

M-203 grenade launcher	2401	2-13
Defensive employment	2402b	2-13 — 2-14
Effect of fire	2406	2-16
Firing positions	2404	2-14
Grip method	Fig. 2-9	2-15
Methods for holding	2404b	2-14
Methods of firing	2405	2-15 — 2-16
Aimed fire	2405a	2-15
Pointing technique	2405b, Fig. 2-10	2-15 — 2-16, 2-16
Offensive employment	2402a	2-13
Trajectory	2403	2-14
Main battle area	5103e	5-4
Minefield:		
Installation	9401d	9-30
Reporting and recording	9401e	9-30
Mine warfare	9401	9-28 — 9-31
Classification of minefields	9401b	9-28 — 9-29
Detection of mines	9401f	9-30 — 9-31
Landmines	9401c	9-29 — 9-30
Minefield installation	9401d	9-30
Purpose of minefields	9401a	9-28
Removal of mines	9401g	9-31
Mission	1001	1-1

N

Net handlers, duties	6004b(4)	6-4
Neutralize, definition	2601a	2-26
Nuclear defense:		
Decontamination	9501e	9-41 — 9-42
Effects on individuals	9501b	9-38 — 9-39
Introduction	9501	9-38 — 9-42
Protection	9501d	9-40 — 9-41
Types of bursts	9501c	9-39 — 9-40

O

Offensive combat:
- Approach march .... Fig. 4-9   4-18
- Conduct phase 4301   4-21
- Exploitation 4401   4-33
- Final preparations:
  - Assembly area 4203a   4-12
  - Attack order .... 4203d   4-14 — 4-16
  - Attack plan 4203c   4-14
  - Troop leading procedures 4203b   4-14
- Infiltration 4601,   4-39,
                                      Fig. 4-11   4-41
  - Assault 4603c   4-42
  - Assembly of groups 4603b   4-42
  - Movement of groups 4603a   4-42
  - Planning and preparation 4602   4-39 — 4-41
- Movement from the assault position through the objective:
  - Assault position .... 4303a   4-25 — 4-26
  - Consolidation 4305   4-30 — 4-31
  - Enemy counterattack 4304   4-29
  - Final coordination line 4303b   4-26
  - Reorganization 4306   4-32
  - Squad in the assault .... 4303c   4-26 — 4-29
- Movement from the line of departure to the assault position:
  - Base of fire element 4302e   4-22
  - Control of the squad .... 4302g   4-22 — 4-23
  - Fire and maneuver 4302a   4-21
  - Fire and movement 4302b   4-21
  - Fire team 4302d   4-22
  - Maneuver element 4302f   4-22
  - Method of advance 4302i   4-24
  - Squad employment 4302c   4-22
  - Use of maneuver .... 4302h   4-23 — 4-24
- Movement to the line of departure:
  - Approach march .... 4204a   4-16 — 4-20
  - Attack position 4204b   4-20
  - Tactical control measures ...... Fig. 4-8   4-17

Night attack:
- Assault 4505b   4-38
- Consolidation and reorganization 4506   4-38

Offensive combat—Continued
    Movement to probable line of deployment....4505a   4-37—4-38
    Preparation..................4504   4-36—4-37
    Purpose and characteristics............4501   4-34
    Security patrols................4503   4-36
    Tactical control measures............4502,   4-34—4-36,
        Fig. 4-10   4-35
  Phases....................4102   4-1
    Conduct..................4102b   4-1
    Exploitation................4102c   4-1
    Preparation................4102a   4-1
  Preparation phase..............4201   4-2
    Movement to the assembly area........4202   4-2—4-12
  Purpose..................4101   4-1
  Special situation..............4205   4-20
Offensive combat preparation, movement
  to the assembly area:
    Connecting elements in a tactical movement....Fig. 4-1   4-3
    Flank patrol................4202b(6)   4-11
    March outpost...............4202b(7)   4-11—4-12
    Point in open terrain.............Fig. 4-2   4-5
    Point of advance guard...........4202b(3)   4-4—4-8
    Point of flank guard............4202b(5)   4-8
    Rear point................4202b(4)   4-8
    Rear point, the squad as..........Fig. 4-5   4-9
    Rear point, withdrawal...........Fig. 4-6   4-10
    Route column...............4202a   4-2
    Sectors of observation, fire team........Fig. 4-3   4-6
    Sectors of observation, individual.......Fig. 4-4   4-7
    Security for halted column..........4202b(7)   4-11—4-12
    Tactical column..............4202b   4-2—4-12
    Tactical column, termination.........4202b(8)   4-12
Organization..................1002   1-1
  Rifle squad, in urbanized terrain.........9105   9-6—9-7

P

Patrol conduct:
  Contact with the enemy............8308   8-25—8-29
  Danger areas................8307   8-25
  Departure and reentry to friendly lines/areas....8302   8-18—8-19

Patrol conduct — Continued
   Exercise of control..................8303   8-20 — 8-21
   Formation and order of movement 8301   8-18
   Immediate action drills...................8308b   8-26
   Movement control measures:
      Checkpoints 8306a   8-23
      Rally points....................8306b   8-23 — 8-24
   Navigation.......................8304   8-21
   Security:
      Day patrols..................8305a   8-22
      Night patrols...................8305b   8-22 — 8-23
Patrol critique 8605   8-44
Patrol organization:
   Combat patrol 8102d   8-2
   Headquarters 8102a   8-1
   Reconnaissance patrol....................8102c   8-1 — 8-2
   Special 8103   8-2
   Task 8104   8-3
   Units 8102b   8-1
Patrol preparations:
   Mission 8201   8-4
   Patrol leader....................8203   8-6
      Complete detailed plans 8212   8-12 — 8-14
      Coordinate 8210   8-11 — 8-12
      Issue patrol order...................8213   8-14 — 8-16
      Issue the warning order 8209   8-9 — 8-11
      Make reconnaissance...................8211   8-12
      Organize the patrol 8207   8-8
      Plan use of time....................8205   8-7
      Select men, weapons, and equipment........8208   8-8 — 8-9
      Study the mission....................8204   8-6
      Study the terrain and situation 8206   8-8
      Supervise, inspect, rehearse, and reinspect 8214   8-16 — 8-17
   Platoon commander's responsibilities...........8202   8-4 — 8-5
Patrol reports 8604,   8-44,
                                                     Fig. 8-3   8-44
Patrolling 8601   8-43
   Captured documents 8603   8-44
   Patrol critique 8605   8-44
   Patrol reports 8604,   8-44,
                                                     Fig. 8-3   8-45
   Sending information 8602   8-43

Plan of defense	5202	5-18 — 5-20
Platoon commander, patrol preparation responsibilities	8202	8-4 — 8-5
Principal direction of fire	5103g, Fig. 5-3	5-5 — 5-7, 5-6
Prisoners of war, handling	App. G	G-1
Probable line of deployment	4502e	4-36

R

Raid patrols	8502a	8-33
Range determination:		
Estimation by eye:		
Appearance of objects	2202b	2-2
Mental unit of measure	2202a, Fig. 2-1	2-2, 2-3
Five-degree method	2203, Fig. 2-2	2-4, 2-5
Importance and methods	2201	2-2
Observation of fire	2204	2-4 — 2-5
Range estimation	App. F	F-1 — F-2
Rate of fire	2605	2-34
Reconnaissance patrols	8401	8-30
Actions at the objective:		
Area reconnaissance	8406a	8-32
Route reconnaissance	8406c	8-32
Zone reconnaissance	8406b	8-32
Equipment	8405	8-31
Missions	8402	8-30
Task organization	8404	8-31
Types of reconnaissance:		
Area	8403a	8-30
Zone	8403b	8-30
Reduced visibility firing:		
Grenade launcher	2604b	2-33
Rifle	2604a	2-33
Rifle and automatic rifle fire	2301	2-6
Beaten zone	2304	2-8
Cone of fire	2303	2-7 — 2-8
Trajectory	2302	2-6

Riflemen:
  Duties 1005f 1-5
  Fire delivery:
    In the attack .... 2603b(2)(a) 2-32
    In the defense .... 2603c(1) 2-33
  Weapons .... 1004a(5) 1-1

S

Sector of fire .... 5103a, 5-2 — 5-3,
                                            Fig. 5-1 5-2
  Forward limit .... 5103a(2) 5-3
  Lateral limits .... 5103a(1) 5-3
Security area 5103f 5-4
Security force, the squad as .... 5210 5-33 — 5-35
Security in defensive combat 5203 5-20
Security patrols:
  Actions at the objective 8507c 8-42
  Mission 8502e 8-34
  Night attack 4503 4-36
  Planning 8507b 8-42
  Purpose 8507 8-41
  Task organization and equipment 8507a 8-42
Shelling reports .... App. F F-1
Squad leader:
  Duties 1005b 1-5
  Position in defensive combat .... 5107e 5-16 — 5-17
  Weapons .... 1004a(1) 1-1
Symbols, military .... App. A A-1
  Colors .... App. A A-1
  Combining .... App. A A-3
  Enemy positions .... App. A A-5
  Miscellaneous .... App. A A-5
  Obstacles .... App. A A-4
  Positions .... App. A A-3
  Sizes .... App. A A-2
  Unit .... App. A A-1 — A-2
  Weapons .... App. A A-4

T

Tank capabilities and limitations .... 9302 9-23 — 9-24
Tank-infantry coordination 9301 9-23
  Infantry teamwork .... 9303 9-24 — 9-25

Tank-infantry tactics:
    Action on contact/attack 9303b   9-25
    Communications 9303c   9-27
    Movement to contact 9303a   9-24
    Safety considerations........................ 9303d   9-27
Tank responsibilities 9301b   9-23
Target assignment 2507   2-21 – 2-23
Target of opportunity, definition 2601c   2-26
Technique of fire 2101   2-1
    Training................................... 2102   2-1
Types of unit fire............................... 2602a   2-26
    Combinations .... 2602d   2-29 – 2-30
    Combinations, engaging two separate targets.... Fig. 2-16   2-30
    Concentrated............................... 2602b,   2-26 – 2-27,
                                         Fig. 2-13   2-27
    Distributed .... 2602c,   2-27 – 2-29,
                                         Fig. 2-14   2-28
    Distributed, engaging two separate targets ...... Fig. 2-15   2-29

U

Unit symbols................................... App. A   A-1 – A-2
Urban terrain 9101   9-1 – 9-2
    Attacking a built-up area 9104   9-6
    Combat techniques 9107   9-9 – 9-12
    Covering party 9105a   9-7
    Defense of a building, preparation............ 9108   9-12 – 9-14
    Rifle squad organization 9105   9-6 – 9-7
    Search party 9105b   9-7
    Search party procedures 9106   9-7 – 9-9
    Structural classification:
        Framed................................. 9102b,   9-3,
                                     Fig. 9-2   9-4
        Frameless............................... 9102a,   9-2 – 9-3,
                                     Fig. 9-1   9-3
    Tactical considerations 9103   9-4 – 9-5

W

Weapons:
  Ammunition:
M-16 rifle	1004b(1)	1-3
M-203 grenade launcher	1004b(3)	1-3 — 1-4
Squad automatic	1004b(2)	1-3
Assault element of fortified attack	9201e	9-19 — 9-20
Assistant automatic riflemen	1004a(4)	1-1
Automatic rifleman	1004a(3)	1-1
Fire team leader/grenadier	1004a(2)	1-1
Organic	1004a	1-1
Rifleman	1004a(5)	1-1
Squad leader	1004a(1)	1-1
Supplementary and munitions	1004c	1-4
Symbols	App. A	A-4

www.ingramcontent.com/pod-product-compliance
Lightning Source LLC
Chambersburg PA
CBHW080331170426
43194CB00014B/2522